OTHER BOOKS BY WILLIAM POUNDSTONE

Big Secrets
The Recursive Universe
Bigger Secrets
Labyrinths of Reason
The Ultimate

PRISONER'S DILEMMA

PRISONER'S DILEMMA

WILLIAM POUNDSTONE

ANCHOR BOOKS
DOUBLEDAY
New York London Toronto Sydney Auckland

AN ANCHOR BOOK
PUBLISHED BY DOUBLEDAY
a division of Bantam Doubleday Dell Publishing Group, Inc.
1540 Broadway, New York, New York 10036

ANCHOR BOOKS, DOUBLEDAY, and the portrayal of an anchor
are trademarks of Doubleday, a division of Bantam Doubleday Dell
Publishing Group, Inc.

Prisoner's Dilemma was originally published in hardcover by
Doubleday in 1992. The Anchor Books edition is
published by arrangement with Doubleday.

Book design by Bonni Leon

Library of Congress Cataloging-in-Publication Data

Poundstone, William.
Prisoner's dilemma / William Poundstone.—1st Anchor Books ed.
p. cm.
Includes bibliographical references and index.
1. Von Neumann, John, 1903–1957. 2. Mathematicians—United States—Biography. I. Title.
QA29.V66P68 1993
510'.092—dc20
[B] 92-29903
CIP

ISBN 0-385-41580-X

Printed in the United States of America
First Anchor Books Edition: February 1993

9 10

TO MIKOLAS CREWS

ACKNOWLEDGMENTS

Many of the original group of game theorists at the RAND Corporation are still well and active. Merrill Flood and Melvin Dresher contributed importantly to this book with their recollections of their work and of the RAND milieu. Much of the biographical material on John von Neumann, including the letters quoted, comes from the collection of von Neumann's papers at the Library of Congress Manuscript Division, Washington, D.C. Some historical information on the Truman administration is based on materials at the Harry S. Truman Presidential Library, Independence, Missouri. Thanks for recollections, assistance, or advice must also go to Paul Armer, Robert Axelrod, Sally Beddow, Raoul Bott, George B. Dantzig, Paul Halmos, Jeane Holiday, Cuthbert Hurd, Martin Shubik, John Tchalenko, Edward Teller, and Nicholas A. Vonneuman.

CONTENTS

PRISONER'S DILEMMA

1

DILEMMAS

A man was crossing a river with his wife and mother. A giraffe appeared on the opposite bank. The man drew his gun on the beast, and the giraffe said, "If you shoot, your mother will die. If you don't shoot, your wife will die." What should the man do?

So asks a traditional "dilemma tale" told by the Popo of Dahomey. Odd and difficult decisions like this are widespread in African folklore. Many have been appropriated by Western writers and philosophers. In the Popo tale, you are supposed to imagine that the pronouncements of talking giraffes are always true. You can restate the dilemma in more Western and technological terms: you, your spouse, and your mother are kidnapped by mad scientists and placed in a room with a strange machine. All three of you are bound immobile to chairs. In front of you is a push button within reach. A machine gun looms in front of your spouse and mother, and a menacing clock ticks away on the wall. One of the scientists announces that if you push the button the mechanism will aim the gun at your mother and shoot her dead. If you *don't* push it within sixty seconds it will aim and fire at your spouse. You have examined the machine and satisfied yourself that its remorseless clockwork will perform as stated. What do you do?

Dilemmas like this are sometimes discussed in college ethics classes. There's no satisfactory answer, of course. It's a cop-out to insist that you should do nothing (don't push the button and allow the machine to kill the spouse) on the grounds that you cannot be "guilty" for doing nothing at all. You can only decide which of the two you like better and spare that one.

Choices are even more difficult when someone else is making a choice, too, and the outcome depends on all the choices made. A similar but more thought-provoking dilemma appears in Gregory Stock's *The Book of Questions* (1987): "You and a person you love deeply are placed in separate rooms with a button next to each of you. You know that both of you will be killed unless one of you presses your button before sixty minutes pass; furthermore, the first person to press the button will save the other person, but will immediately be killed. What do you think you would do?"

Now there are two people pondering their predicament and making independent choices. It's vital that *someone* push the button. The tricky part is *when* you should make the sacrifice. The dilemma forces you to make a "lifeboat" decision about you and your loved one. Who really should be the one to live?

There are many situations where one person might elect to save the other at the sacrifice of his own life. A parent might save a child on the grounds that the child is likely to have more years of life left. Whatever the criteria used—and there is no reason to believe both persons would use the same criteria—there are three possible outcomes of the lifeboat decision.

The least disturbing case is when both come to the same conclusion about who should make the sacrifice and who should be saved. Then the former should push the button and save the latter.

A second possibility is that both persons decide to save each other. A mother decides to save her daughter, who has more years left, and the daughter decides to save the mother who gave her life. In that case there is a race to push the button first.

The most disturbing case is when *both* persons decide *he* or *she* is the one who should be saved. Then neither pushes the button and the clock ticks away . . .

Think about this third scenario a bit. The clock reads fifty-nine minutes after the hour. You haven't pushed the button, hoping that your loved one would—only he/she hasn't. (Let's assume that the survivor is notified immediately when the other person pushes the button.) You've had time to mull over the possibilities. Some people might take the whole hour to decide who to save, or to work up the courage to push the button. But it's beginning to look as if your loved one has decided *you* should make the sacrifice.

There's no point in vowing *never* push the button, not even in the very last split second. No matter how self-centered you may be, you have no power to save yourself. Someone has to die; that's the way the dilemma works. If your loved one fails to make the sacrifice, you might as well save him or her. Remember, you really love the other person.

Ideally, you would want to push the button in the last instant. It is possible that your loved one is thinking the very same thing. That gives you all the more reason to hold off until the very, *very* last instant. You want to give the other person a chance to push the button in the "last instant," but if he/she doesn't, then and only then do *you* push the button. Of course, your loved one may be planning *this* as well.

Complicating this mutual intention to hold off until the very last moment are matters of reaction times and clock accuracy. The infernal machine has no sympathy for an *intention* to have pushed the button within the time limit. You or the other person must physically complete a circuit before the machine overrides it and kills both of you. It is unclear whether the clock on the wall is *perfectly* synchronized with the machine. The mad scientist said it was, but he may not have been talking about the second hand, and he *is* mad. To be sure of pushing the button within the time limit, you really ought to push it a moment or two early, just to be safe. That's an awkward requirement when applying the wait-until-the-very-last-moment strategy!

Your procrastinating loved one is in the same boat. Should both of you resolve to push the button only at the last instant, the outcome would be a crapshoot. One person would happen to push it a split second before the other, or both would miss the deadline and both would die. The outcome would in effect be determined randomly. Chance would triumph over rationality.

Desperate decisions in strange rooms have enough currency in the philosophical literature to have earned a name: "problem boxes." Why are such dilemmas so compelling? It's partly the novelty of bizarre predicaments. But they would hardly inspire such interest if they were just puzzles with no resonance in our personal experience.

The dilemmas of real life are created not by mad scientists but by the sundry ways that our individual interests clash with those of others and of society. We daily face difficult decisions, and sometimes our choices work out differently from the way we expected. The inner question raised by dilemmas is simple and troubling: Is there a rational course of action for every situation?

THE NUCLEAR DILEMMA

In August 1949, the Soviet Union exploded its first atomic bomb in Siberia. The U.S. nuclear monopoly was over. Much sooner than Western observers had expected, there were two atomic powers.

The Soviet bomb sparked a nuclear arms race. Some of the consequences of that race were easy to predict. Each nation would arm to the point where it was capable of launching a swift, devastating nuclear attack on the other. That, many thought, would pose an unacceptable dilemma. For the first time in the history of the world, it became possible to contemplate a surprise attack that would wipe an enemy nation off the face of the earth. In times of crisis, the temptation to push the nuclear button would be irresistible. Equally important, each nation would fear being the *victim* of the other's surprise attack.

By 1950, a number of people in the United States and Western Europe had decided that the United States should contemplate an immediate, unprovoked nuclear attack on the Soviet Union. This idea, which went by the euphemistic name of "preventive war," held that America should seize the moment and establish a world government through nuclear blackmail or surprise attack. You might think that only a lunatic fringe would support such a plan. In fact, the preventive war movement found support among many of undeniable intelligence, including two of the most brilliant mathematicians of the time: Bertrand Russell and John von Neumann. Mathematicians are not usually known for their political opinions, or for worldly views of any kind. In most respects Russell and von Neumann were quite dissimilar men. But they concurred that there wasn't room for two atomic powers in the world.

Russell, one of the mainsprings of the preventive war movement, spoke in favor of an ultimatum threatening the Soviet Union with nuclear devastation unless it surrendered sovereignty to a U.S.-dominated world state. In a 1947 speech, Russell said, "I am inclined to think that Russia would acquiesce; if not, provided this is done soon, the world might survive the resulting war and emerge with a single government such as the world needs."

Von Neumann took a harder line yet, favoring a surprise nuclear first-strike. *Life* magazine quoted von Neumann as saying, "If you say why not bomb them tomorrow, I say why not today? If you say today at 5 o'clock, I say why not one o'clock?"

Neither man had any love of the Soviet Union. But they believed preventive war was foremost a matter of logic, the only rational solution to the deadly dilemma of nuclear proliferation. As Russell put it in an article advocating preventive war in the January 1948 issue of *New Commonwealth:* "The argument that I have been developing is as simple and as unescapable as a mathematical demonstration." But logic itself can go awry. Nothing captures the whole bizarre episode of preventive war better than the unintentionally Orwellian words of

U.S. Secretary of Navy Francis P. Matthews, who in 1950 urged the nation to become "aggressors for peace."

Today, with East-West tensions relaxing, preventive war seems a curious aberration of cold-war mentality. Yet the same sorts of issues are very much with us today. What should a nation do when its security conflicts with the good of all humanity? What should a person do when his or her interests conflict with the common good?

JOHN VON NEUMANN

Perhaps no one exemplifies the agonizing dilemma of the bomb better than John von Neumann (1903–1957). That name does not mean much to most people. The celebrity mathematician is almost a nonexistent species. Those few laypersons who recognize the name are most likely to place him as a pioneer of the electronic digital computer, or as one of the crowd of scientific luminaries who worked on the Manhattan Project. A few may recognize him as one of several alleged models for "Dr. Strangelove" of the 1963 Stanley Kubrick film. It was, at any rate, von Neumann who attended Atomic Energy Commission meetings in a wheelchair.

The main body of von Neumann's work, the work that early made his reputation for genius, lies in inaccessible realms of pure mathematics and mathematical physics. It might have been expected that all his life's work would be remote from the affairs of the world. Von Neumann, however, had a passion for applied mathematics. Both the computer and the bomb were extracurricular projects for von Neumann, but they typify his interest in the applications of mathematics.

Von Neumann was a poker player—not an especially good one. His nimble mind picked up on certain elements of the game. He was particularly interested in the deception, the bluffing and second-guessing, the way players try to mislead each other within the framework of the rules. There was, in mathematical jargon, something "nontrivial" about that.

From the mid 1920s through the 1940s, von Neumann amused himself with an investigation into the mathematical structure of poker and other games. As this work took shape, he realized that its theorems could be applied to economics, politics, foreign policy, and other diverse spheres. Von Neumann and Princeton economist Oskar Morgenstern published their analysis in 1944 as *Theory of Games and Economic Behavior.*

The first thing to realize about von Neumann's "game theory" is that it is only tangentially about games in the usual sense. The word "strategy," as commonly used, better suggests what game theory is about. In *The Ascent of Man,* scientist Jacob Bronowski (who worked with von Neumann during World War II) recalls von Neumann's explanation during a taxi ride in London:

> *. . . I naturally said to him, since I am an enthusiastic chess player, "You mean, the theory of games like chess." "No, no," he said. "Chess is not a game. Chess is a well-defined form of computation. You may not be able to work out the answers, but in theory there must be a solution, a right procedure in any position. Now real games," he said, "are not like that at all. Real life is not like that. Real life consists of bluffing, of little tactics of deception, of asking yourself what is the other man going to think I mean to do. And that is what games are about in my theory."*

Game theory is a study of conflict between thoughtful and potentially deceitful opponents. This may make it sound like game theory is a branch of psychology rather than mathematics. Not so: because the players are assumed to be perfectly rational, game theory admits of precise analysis. Game theory is therefore a rigorous branch of mathematical logic that underlies real conflicts among (*not* always rational) humans.

Most great advances in science come when a person of insight recognizes common elements in seemingly unrelated contexts. This describes the genesis of game theory. Von Neumann recognized that parlor games pose elemental conflicts. It was these conflicts, normally obscured by the window dressing of cards and chessmen and dice, that occupied von Neumann. He perceived similar conflicts in economics, politics, daily life, and war.

As von Neumann used the term, a "game" is a conflict situation where one must make a choice knowing that others are making choices too, and the outcome of the conflict will be determined in some prescribed way by all the choices made. Some games are simple. Others invite vicious circles of second-guessing that are difficult to analyze. Von Neumann wondered if there is always a rational way to play a game, particularly one with much bluffing and second-guessing. This is one of the fundamental questions of game theory.

You might naively suppose that there must be a rational way to play every game. *Is this necessarily so?* von Neumann wanted to know. The world is not always a logical place; much of the irrational fills our daily lives. More to the point, the mutual second-guessing in games like poker invokes potentially infinite chains of reasoning. It was not obvious that rational players would ever come to a definite conclusion about how to play.

A less talented mathematician might have posed the same questions, sighed, and gone back to "serious" work. Von Neumann tackled the problem head-on with full mathematical rigor. He came up with a remarkable proof.

Von Neumann demonstrated mathematically that there is always a rational course of action for games of two players, *provided their interests are completely opposed.* This proof is called the "minimax theorem." The class of games covered by the minimax theorem includes many recreational games, ranging from such trivial contests as ticktacktoe to the sophistications of chess. It applies to any two-person game where one person wins and the other loses (this is the simplest way of meeting the requirement that the players' interests are "completely opposed"). Von Neumann proved that there is always a "right," or more exactly, an "optimal," way to play such games.

Were this all there was to the minimax theorem, it would qualify as a clever contribution to recreational mathematics. Von Neumann saw more profound implications. He intended the minimax theorem to be the cornerstone of a game theory that would eventually encompass other types of games, including those of more than two players and those where the players' interests partly overlap. So expanded, game theory could encompass any type of human conflict.

Von Neumann and Morgenstern presented their game theory as a mathematical foundation for economics. Economic conflicts can be viewed as "games" subject to the theorems of game theory. Two contractors submitting bids for a contract, or a group of buyers bidding at an auction, are enmeshed in subtle games of second-guessing that are open to rigorous analysis.

Game theory was hailed as an important new field almost from the beginning. A review of von Neumann and Morgenstern's book in the *American Mathematical Society Bulletin* predicted, "Posterity may regard this book as one of the major scientific achievements of the first half of the twentieth century. This will undoubtedly be the case if the authors have succeeded in establishing a new exact science—the science of economics." In the years after the publication of *Theory of Games and Economic Behavior,* game theory and its terms became popular buzzwords with economists, social scientists, and military strategists.

One of the places where game theory found immediate acceptance was the RAND Corporation. RAND, the prototypic "think tank," was founded at the Air Force's behest shortly after World War II. The RAND Corporation's original purpose was to perform strategic studies on intercontinental nuclear war. RAND hired many of the scientists leaving wartime defense work, and took on as consultants an even larger orbit of stellar thinkers.

RAND thought highly enough of game theory to hire von Neumann as a consultant and to devote a great deal of effort not only to military applications of game theory but also to basic research in the field. During the late 1940s and early 1950s, von Neumann was a regular visitor to RAND's Santa Monica, California, headquarters.

PRISONER'S DILEMMA

In 1950 two RAND scientists made what is arguably the most influential discovery in game theory since its inception. Merrill Flood and Melvin Dresher devised a simple, baffling "game" that challenged part of the theoretical basis of game theory. RAND consultant Albert W. Tucker dubbed this game the *prisoner's dilemma.* So-called because of a story Tucker told to illustrate it, the game was quickly recognized as the most chimerical of conflict situations. We'll come to the prisoner's dilemma in due course. For now, it will suffice to say it is an intellectual riddle that still puzzles us today.

To those who study folklore, a dilemma tale is a story that presents a difficult decision and asks the listener what he would do. The prisoner's dilemma is just that, a story ending in an unresolved dilemma left to the listener or reader. Not published as such until years after its invention, the prisoner's dilemma spread through the scientific community of the 1950s by oral transmission that would satisfy the folklorist's definition of a dilemma tale.

Of course, the prisoner's dilemma is much more than a story. It is a precise mathematical construct and also a real-life problem. By the sort of synchronicity that is less mysterious than it seems, the prisoner's dilemma was "discovered" in 1950, just as nuclear proliferation and arms races became serious concerns. Today, the tensions of the early nuclear era are considered the classic illustration of a prisoner's dilemma.

The troubling question of whether to gain security for one side at the expense of the common good is not new to the nuclear age. Such issues are as old as warfare. But the swiftness and devastation of nuclear attack brought these issues to the fore. It is not much exaggeration to say that the prisoner's dilemma is the central issue of defense, and that one's personal reaction to it (which cannot be proven right or wrong) is what makes some people conservatives and other people liberals.

This is not a book on military strategy. The prisoner's dilemma is a universal concept. Theorists now realize that prisoner's dilemmas occur in biology, psychology, sociology, economics, and law. The prisoner's dilemma is apt to turn up anywhere a conflict of interests exists —and the conflict need not be between sentient beings. Study of the prisoner's dilemma has great power for explaining why animal and human societies are organized as they are. It is one of the great ideas of the twentieth century, simple enough for anyone to grasp and of fundamental importance.

In the last years of his life, von Neumann saw the realities of war becoming more like a fictional dilemma or the abstract games of his theory. The perils of the nuclear age are often attributed to "technical progress outstripping ethical progress." This diagnosis is all the more disheartening for the suspicion that there is no such thing as ethical progress, that bombs get bigger and people stay the same. The prisoner's dilemma has become one of the premier philosophical and scientific issues of our time. It is tied to our very survival.

Today's practitioners of game theory are attempting to forge a kind of ethical progress. Is there any way to promote the common good in a prisoner's dilemma? The attempt to answer this question is one of the great intellectual adventures of our time.

Atomic Energy Commissioner, an editor of the *Jewish Daily Forward* wrote von Neumann to ask if it was correct to describe him as Jewish.

Max Neumann was cultured and well read. He was an amateur poet in both Hungarian and German. At work he was an enlightened businessman who felt it fitting to consider the social desirability of businesses he was financing. Dinner-table discussions with his children often touched on the social responsibilities of bankers. He tried to interest his sons in banking by bringing home mementos of the businesses his firm was backing. Von Neumann's brother Nicholas speculated that Johnny's idea for computer punched cards was inspired by the family's discussion of the Jacquard loom factory that Max's bank financed.

In 1913 Max purchased a *margittai* title of nobility, which the cash-strapped Emperor Franz Joseph was then retailing to the rising capitalist class. Many changed their names upon receiving the title, but Max did not except to prefix the title: "margittai Neumann." During his Zurich and Berlin university days, John affected the Germanized version, signing himself "Johann Neumann von Margitta." The surname elided, in accord with German usage, to the shorter "von Neumann," where the title is not mentioned but is aristocratically implied.

THE CHILD PRODIGY

From childhood, von Neumann was gifted with a photographic memory. At the age of six, he was able to exchange jokes with his father in classical Greek. The Neumann family sometimes entertained guests with demonstrations of Johnny's ability to memorize phone books. A guest would select a page and column of the phone book at random. Young Johnny read the column over a few times, then handed the book back to the guest. He could answer any questions put to him (who has number such and such?) or recite names, addresses, and numbers in order.

The Neumann household was a congenial environment for a child prodigy's intellectual development. Max Neumann bought a library in an estate sale, cleared one room of furniture to house it, and commissioned a cabinetmaker to fit the room with floor-to-ceiling bookcases. Johnny spent many hours reading books from this library. One was the encyclopedic history of the world edited by the once-fashionable German historian Wilhelm Oncken. Von Neumann read it volume by

2

JOHN VON NEUMANN

John von Neumann was born in Budapest, Hungary, on December 28, 1903. "John" is anglicized from János, his given name. In Hungary, he was called by the diminutive Jancsi, which changed naturally to Johnny in the United States. Johnny was the oldest of three sons of Max Neumann, a successful banker. The Neumann family lived in a comfortable three-story apartment house owned by von Neumann's maternal grandfather, Jakab Kann. Four of Kann's daughters and their families lived in the building, so the other children were von Neumann's cousins. The extended family had both German and French governesses to help the children acquire the linguistic fluency deemed necessary for success in Hungarian society. After the early 1920s, summers were spent in another house Max Neumann bought in the suburbs of Budapest.

The Neumanns were Jews, persecuted but often prosperous outsiders in the society of Magyar Hungary. In *The Ascent of Man,* Jacob Bronowski asserts improbably of von Neumann, "If he had been born a hundred years earlier, we would never have heard of him. He would have been doing what his father and grandfather did, making rabbinical comments on dogma."

In fact the family's religious attitudes were ambivalent. Patriarch Jakab Kann's piety was not inherited by most of those living under his roof. Max Neumann's family were so ecumenical that they put up a Christmas tree and exchanged gifts each holiday season. The children sang Christmas carols with their German governess. This did not preclude equally secular observance of the major Jewish holidays. Once Johnny's brother Michael asked his father why the family considered itself Jewish when it did not observe the religion seriously. Max answered: "Tradition."

This religious confusion would follow von Neumann through a life that would include a nominal conversion to Catholicism at the time of his first marriage and essentially agnostic beliefs for most of his adult life. Near the end of his life, at the height of his public renown as an volume. He would balk at getting a haircut unless his mother let him take a volume of Oncken along. By the outbreak of World War I, Johnny had read the entire set and could draw analogies between current events and historical ones, and discuss both in relation to theories of military and political strategy. He was ten years old.

Von Neumann was exposed to psychology through a relative, Sándor Ferenczi, a disciple of Freud who had introduced psychoanalysis into Hungary. Von Neumann also had important exposure to European literature and music at an early age. His brother Nicholas recalls that Johnny was intrigued by the philosophical underpinnings of artistic works. Pirandello's *Six Characters in Search of an Author* appealed to Johnny for its confusion of reality and make-believe. Bach's "Art of the Fugue" left an impression due to the fact that it was written for several voices with the instruments unspecified. Johnny was so impressed with this that Nicholas credits it as a source for the idea of the stored-program computer.

An early interest in science led him to conduct home experiments. Johnny and his brother Michael somehow got a piece of sodium and dropped it into water to watch the reaction. After the sodium had dissolved (producing caustic sodium hydroxide), they tasted the water. The worried family contacted a physician.

From 1911 through 1921 von Neumann attended the Lutheran Gymnasium for boys, a high school with a strong academic reputation. Despite its religious affiliation, the gymnasium accepted students of all backgrounds and even provided appropriate religious training. Von Neumann's first math teacher at the gymnasium, Professor Laszlo Rácz, quickly recognized his talent and called Max in for a conference. Rácz recommended a special math study program for von Neumann and set about organizing it.

Von Neumann was occasionally an exasperation to his teachers. He would confess that he had not studied the day's assignment, then participate in the discussion more knowledgeably than anyone else. Johnny got straight A's in math and most of the academic subjects. He got C's in physical education.

The Hungary of von Neumann's generation produced an extraordinary number of geniuses. Von Neumann was a schoolmate of several. He was a year behind Eugene Wigner, later a renowned physicist, at the gymnasium. (In adulthood, Wigner said he realized he would be a second-rate mathematician compared with von Neumann and therefore turned to physics.) Johnny met Edward Teller in 1925 when both studied under a renowned teacher, Lipot Feher. Other notable Hungarians of the time included laser pioneer Dennis Gabor and physicist-turned-biologist Leo Szilard.

KUN'S HUNGARY

Despite a stable home life, a nurturing intellectual environment, and relative insulation from the horrors of the World War, von Neumann did not reach adulthood without a taste of hardship and persecution. The Neumanns' experience of the Kun regime deserves mention, as it may bear on von Neumann's political conservatism and distrust of the Soviet Union.

Béla Kun was a utopian socialist who headed a disastrous five-month-long Communist government in Hungary. Of Jewish Hungarian stock, Kun was an attorney and journalist. He was drafted into the Austro-Hungarian Army and captured by the Russians. One story claims he was brainwashed in prison. Kun had socialist convictions before the war, and it is unclear to what degree his politics were shaped by the prison experience. In any event, Kun returned to his native land an ardent Communist. He organized a Communist party in Budapest after Soviet models. The party grew quickly.

Meanwhile, Count Mihály Károlyi's government foundered. In a last-ditch attempt to preserve power, Károlyi appealed to Kun's powerful Communist party for support. This only eroded Károlyi's position further with his conservative constituency. On March 21, 1919, Károlyi accepted the inevitable and resigned. Kun seized power as head of a workers and peasants' state. Johnny von Neumann was fifteen at the time.

Kun lacked leadership skills and, at times, good judgment. He was a doctrinaire socialist who applied the prescriptions of Marx and Lenin literally. When visionary acts failed to produce a promised utopia, Kun seemed unable to do much more than repeat revolutionary slogans. Under Kun, Hungary became a compendium of mismanagement.

One of Kun's first acts was to issue a decree transferring ownership of land, businesses, and means of production to the proletariat. This threw the centuries-old ruling classes from power overnight. Important jobs were filled with inexperienced socialists or opportunists. It was claimed that the new commissar of finances had to be shown how to endorse a check: he had never done it before. This story may be apocryphal, but it illustrates the tragicomic predicament.

The economy ground to a halt. The poor found little incentive to work for the state. Store shelves went bare as farmers refused to sell their goods to the state. To get food, city dwellers had to hike to the countryside and bargain with farmers. Those who had something to sell demanded the outlawed blue money rather than the Kun regime's white money.

Bank officers were locked out. The affluent who still had jobs were often afraid to report for work. Max Neumann fled with his family to Austria about a month after Kun came to power. The Neumanns' exile was as comfortable as could be expected. They divided their time between Vienna and the Adriatic resort of Abbazia.

Budapest became a dangerous place. Youths called "Lenin Boys" used the upheaval as an excuse for violence and vandalism. They roamed neighborhoods of the wealthy, knocking on doors and beating those who answered. Pockets of conservative resistance added to the tension. "Traitors" were arrested and sometimes shot.

Kun's government collapsed in August 1919. Admiral Miklós Horthy seized power with the support of the displaced aristocracy and radical military factions of the stripe that would soon be called fascist. Hungary's troubles were not over. Under the conservative Horthy regime, a "White Terror" gripped Hungary.

Eight of Kun's eleven commissars had been Jews. The failure of the Kun government was quickly blamed on Jews and intellectuals. The White Terror was unreasoning, attacking even those Jews, like the Neumanns, who had opposed and been victimized by Kun's rule. About 5,000 people were killed, often by lynch mobs who operated without government interference. It is estimated that about 100,000 people fled Hungary.

EARLY CAREER

John von Neumann became part of an exodus of Jewish intellectuals who left Hungary for Germany and then Germany for the United States. As Johnny approached college age, he wanted to study mathematics. He did in fact publish his first mathematical paper in collaboration with his tutor at the age of eighteen. His father disapproved, believing there was not enough money in being a mathematician. Max asked engineer Theodore von Kármán to talk to his son and convince him to choose a business career. Von Kármán suggested chemistry as a compromise, and both father and son agreed.

Johnny applied to the University of Budapest. State quotas limited admissions of Jews to their proportion in the population, a rule that made the academic requirements for the relatively well-educated Jews anomalously high. Of course, with von Neumann's brilliant record, this was no problem, and he was admitted.

Von Neumann's complex college career spanned three nations, however. In 1921 he enrolled in the University of Budapest but did not attend classes. He showed up only to ace the exams. Simultaneously, he enrolled at the University of Berlin, where he studied chemistry through 1923. After Berlin, his academic grand tour took him to the Swiss Federal Institute of Technology in Zurich. There he studied chemical engineering, earning a degree in 1925. Finally, he received his Ph.D. in mathematics—with minors in physics and chemistry—from the University of Budapest in 1926. It had taken just five years to earn the Ph.D.

After taking his Ph.D., he was named *Privat dozent* at the University of Berlin. A *privat dozent* is comparable to an assistant professor in the United States, and von Neumann was reportedly the youngest man ever to hold that position. He remained at the Berlin post through 1929. Then he went to Hamburg, holding the same title through 1930.

Simultaneously, von Neumann received a Rockefeller grant for postdoctoral work at the University of Göttingen. There he studied (1926–27) under the great mathematician David Hilbert, who had gathered many of the most promising mathematical minds around him. One was J. Robert Oppenheimer, whom von Neumann met for the first time.

By his mid-twenties, von Neumann's fame had spread worldwide in the mathematical community. At academic conferences, he would find himself pointed out as a young genius. The young von Neumann brashly asserted that mathematical powers declined after the age of twenty-six; only the benefits of experience concealed the decline—for a time, anyway. (His longtime friend, mathematician Stanislaw Ulam, reported that as von Neumann himself aged, he raised this limiting age.)

In 1929, the year von Neumann reached this supposed watershed, he was invited to lecture on quantum theory for a semester at Princeton. Upon being offered the job, he resolved to marry his girlfriend, Mariette Koevesi. He wrote back to Oswald Veblen of Princeton that he had to attend to some personal matters before he could accept. Von Neumann returned to Budapest and popped the question.

His fiancée, daughter of a Budapest doctor, agreed to marry him in December. Mariette was Catholic. Von Neumann accepted his wife's faith for the marriage. Most evidence indicates that he did not take this "conversion" very seriously. In Stanley A. Blumberg and Gwinn Owens' *Energy and Conflict,* Edward Teller said that whenever von Neumann was tempted to curse, he would restrain himself and joke, "Now I will have to spend two hundred fewer years in purgatory."

Von Neumann accepted the Princeton appointment. His duties at Princeton were soon upgraded to a permanent position. His short teaching career dates from 1930 to 1933, when he held the title of Visiting Professor. Von Neumann was said to be an indifferent teacher. His fluid line of thought was difficult for those less gifted to follow. He was notorious for dashing out equations on a small portion of the available blackboard and erasing expressions before students could copy them.

From 1930 through 1936, von Neumann spent summers in Europe on working vacations. Not until 1933, the year the Nazis came to power, did von Neumann finally give up his professional affiliations in Germany. As would be expected, von Neumann opposed the Nazis from their emergence. In 1933 he made the often-quoted prediction (in a letter to Veblen) that "if these boys continue for only two more years (which is unfortunately very probable), they will ruin German science for a generation—at least."

The role of anti-Semitism in von Neumann's migrations should not be overemphasized. In later life von Neumann insisted that both the move from Hungary to Germany and from Germany to the United States was motivated by career opportunities. There were too many mathematicians and not enough full professorships in Germany, he complained.

THE INSTITUTE

When the Princeton Institute for Advanced Study opened in 1933, von Neumann was named a professor. Again, he was the youngest of a very distinguished group. As the institute was not a teaching institution, this marked the end of von Neumann's formal teaching career. (Mercifully, perhaps, he had supervised just one Ph.D. thesis.)

The institute was headquartered in an unimpressive building often compared to a Howard Johnson's. Most new members did, and still do, go through a "shock of recognition" phase in which they find that the most ordinary-looking of people are famous (to professionals) figures. "We had to pinch ourselves at times to be certain that it was all real," mathematician Raoul Bott recalled in a 1984 speech. "Imagine a place where the suspicious-looking vagrant whom the police try to arrest turns out to be Jean Leray; where around eleven each morning it is quite easy to chat with Einstein about weighty subjects such as the weather or the tardiness of mail delivery. Where the friendly but very silent neighbor in the midst of a raucous group of young lunchers turns out to be P. A. M. Dirac, and so on and so on."

Von Neumann had an office near Einstein's. Though popular articles on von Neumann, struggling to explain just who he was, sometimes painted him as a "collaborator" of the higher-profile genius, the two were never close. Von Neumann's brother Nicholas told me that Johnny viewed Einstein's work on the "unified field theory" with indulgent (and accurate) skepticism. Einstein had been twenty-six (!) when he published the special theory of relativity. (Einstein still had a good decade—at least—after that, for the general theory of relativity came eleven years later.) *Life* magazine quoted a member of the institute: "Einstein's mind was slow and contemplative. He would think about something for years. Johnny's mind was just the opposite. It was lightning quick—stunningly fast. If you gave him a problem he either solved it right away or not at all. If he had to think about it a long time and it bored him, his interest would begin to wander. And Johnny's mind would not shine unless whatever he was working on had his undivided attention."

Mariette gave birth to a daughter, Marina, in 1935. Marina was von Neumann's only child. Two years later, Mariette left Johnny. She had fallen in love with J. B. Kuper, a physicist.

The evidence suggests that the von Neumanns had so completely grown apart as to permit a graceful and civilized separation. Kuper, who worked at Brookhaven National Laboratory, would be described as a "very nice boy" and "terribly in love" with Mariette by friends of von Neumann the following year. Mariette moved in with Kuper, taking Marina. Late in 1937, she took up residence in Reno for a Nevada divorce. She filed on grounds of "extreme cruelty." This seems to be a legal fiction, for Johnny remained on good terms with Mariette afterward. He called her almost daily during a 1938 trip to Europe. Once the divorce was final, Mariette married Kuper and settled in New York City.

The separation agreement gave Mariette no alimony, only child support for Marina. It provided for Marina to live with her mother only up to the age of twelve. Then she would live with her father through age eighteen. "My mother felt very strongly that any kid who was John von Neumann's daughter should get to know John von Neumann," Marina said in a 1972 *Life* magazine interview. "It was very carefully worked out that I would live with him in my teens when I was old enough to reap the advantages."

KLARA

Within a year of his divorce, von Neumann was in love with a married woman. He rekindled a relationship with a childhood sweetheart, Klara Dan, during a trip to Europe.

Like von Neumann, Klara (also called Klari) came from a comfortably affluent Budapest family. In a 1956 interview, Klara described herself as having been "a spoiled brat, in luxurious European style. I lived for fun, in England and on the Riviera. I had a marvelous time, but I actually did not know the meaning of the word 'science.' "

Klara was willing to divorce her husband. Von Neumann took off the fall 1938 semester from the institute, ostensibly to assist with Klara's divorce. Actually, he had a busy agenda of lectures. Von Neumann defied the gathering clouds of war to lecture in Sweden and Poland and call on several "clever friends" (as Klara called them), including the more-than-clever physicist Niels Bohr in Copenhagen. This left relatively little time to spend with Klara and family in Budapest.

Von Neumann's letters, saved by Klara and now archived among his papers at the Library of Congress, chronicle a passionate and stormy relationship. The letters are numerous by mid-twentieth-century standards, for von Neumann traveled extensively not only during those months in Europe but throughout the marriage.

During their courtship, Johnny wrote Klara exuberant love letters. He lavishes praise on Klara as the finest woman in creation and declares their love to be a sensation without parallel. This lyrical tone is almost entirely absent from Klara's letters to Johnny. Klara was a shrew. It is difficult to come to any other conclusion from reading her letters. Even the earliest letters, written in the very springtime of their love, abound with threats and ominous undertones. If John von Neumann wasn't the easiest person in the world to live with, neither was Klara.

In a letter written at the Grand Hotel, Montecatini Terme, Italy, dated August 28, 1938, Klara describes herself frankly as a vain woman with many faults. She values stability in a suitor. In the same letter she chides Johnny, as she was to do in letter after letter for much of their lives. She carps that Johnny is immature, a frightened boy who expects to get his way because of his genius. She charges that he cares nothing for her feelings or the feelings of others.

Klara was subject to persistent moodiness. A group of letters written in September 1938 give some idea: On September 9, Klara complains of feeling rotten; on September 15, she can't even bear to meet people and make small talk—everyone else is disgustingly optimistic. On September 19, she complains of depression again and sees no reason why she will ever get better.

Incredibly, one of Johnny's letters from this time has a plummy P.S. insisting that every letter of Klara's is twice as nice as the preceding one.

Klara had ample reason to be anxious. With war and the horrors of the Nazis and fascists looming, she found in von Neumann not only a potential husband but a ticket to America. In a heartfelt letter, Klara tells Johnny that her very survival may be in his hands. All she wants is to go to America and lead a normal married life. After several agonizing (to Klara) delays, Klara's divorce was granted October 29, 1938. She and Johnny married on November 17. Shortly afterward, they set sail on the *Queen Mary* for New York.

The von Neumanns lived at 26 Westcott Road in one of Princeton's biggest houses. The house was valued at $30,000—a mortgage balance of $25,000 was owed to the Institute for Advanced Study, which financed home purchases for the faculty. (In 1941 von Neumann was earning $12,500 a year from the institute, a good academic salary for the time.)

Klara decorated the house with modern furniture, a style to which Johnny was indifferent. In other respects, the von Neumann household had a distinctly Old World flavor. Two, then three, generations lived under one roof (von Neumann's mother lived with them, and

Marina moved in when she turned twelve). There were servants as well. Johnny and Klara mostly spoke Hungarian between themselves. Their letters are in both Hungarian and English, sometimes switching languages in the middle of a sentence.

Klara and Johnny did not have any children together. They were attentive parents to Marina, and they doted on their large mixed-breed dog, "Inverse."

PERSONALITY

"There is no scientific 'queerness' about Dr. von Neumann's appearance," declared *Good Housekeeping* magazine in 1956. Von Neumann stood five feet nine inches tall. Never athletic, he was slender in his twenties, becoming plump as middle age set in. His hairline had receded to the top of his head, creating a moon-faced appearance. An Alfred Eisenstaedt photograph of him at a Princeton tea, taken for *Life* magazine, suggests a kindly milquetoast uncle. He dressed neatly and conservatively. Klara went with Johnny to help him pick out his clothes, claiming he had low sales resistance. He usually wore prim, vested suits with a white handkerchief in the pocket, an outfit just enough out of place to inspire pleasantries. "Why don't you get some chalk dust on your clothes so you'll look like the rest of us?" colleagues asked. It was claimed that he carried little in his pockets except security clearances and Chinese puzzles.

Von Neumann's mastery of English was excellent. For that matter, so was his mastery of Hungarian, German, and French. His English betrayed a Middle European accent that was invariably described as charming, never harsh. He had trouble pronouncing "th" and "r," and pronounced "integer" with a hard *g*—this being a von Neumann trademark. He retained a strong grasp of both Greek and Latin learned in childhood. It was said that von Neumann could speak in any of seven languages faster than most people could speak in one.

The stereotype of a mathematician is someone shy and vague. Von Neumann was a heartier sort: literally, the life of the party. Parties and nightlife held a special appeal for von Neumann. While teaching in Germany, von Neumann had been a denizen of the *Cabaret*-era Berlin nightlife circuit. This enthusiasm continued throughout his life. Johnny and Klara took it upon themselves to introduce guest scholars to the institute faculty. They had parties every week or so, creating a kind of salon. In 1952, Klara had an ice sculpture of the Princeton computer for a party celebrating its inauguration. *Life* magazine wrote of von Neumann: "It was a common occurrence for him to begin scribbling with pencil and paper in the midst of a night-club floor show or a lively party, 'the noisier,' his wife says, 'the better.'"

Von Neumann used his phenomenal memory to build a mental library of jokes he used as the occasion demanded. In 1947, Atomic Energy Commissioner Lewis Strauss wanted to spice up a speech and asked von Neumann for several jokes about the atomic bomb (sadly, not recorded). Von Neumann also enjoyed limericks. In letters to Klara, he included these examples:

There was a young lady from Lynn
Who thought that to love was a sin,
But when she was tight
It seemed quite alright,
So everyone filled her with gin!

There was a young man who said: Run!
The end of the world has begun!
The one I fear most
Is that damn' Holy Ghost,
I can handle the Father and Son.
(Unknown limerist, early XXIst Century)

Much of von Neumann's humor is hopelessly sexist by contemporary standards. In a letter to Klara, he defined "rape," *Playboy*'s Unabashed Dictionary-style, as "assault with intent to please." Slightly more clever is his story of a woman who tries to spend a five-dollar bill in a store and is told it is counterfeit. "I've been raped!" the woman says. A clue to Klara's reaction to her husband's jokes occurs in a letter where Johnny quotes P. G. Wodehouse: "Women have to learn to bear anecdotes from men they love. It is the curse of Eve."

Some of von Neumann's political humor bears up better. Lewis Strauss attributed to von Neumann the line, "It takes a Hungarian to go into a revolving door behind you and come out first." One of von Neumann's favorite jokes was set in Berlin during World War I. A man stands on a corner and yells, "The Kaiser is an idiot!" Police come and arrest him for treason. The man says, "You can't arrest me, I was talking about the Austrian Kaiser." The police say, "You can't fool us, we know who the idiot is."

Von Neumann was also known for practical jokes. Merrill Flood recalls the time that Einstein was supposed to go to New York and von Neumann offered to drive him to the Princeton train station. Von Neumann purposely put Einstein on a train going in the wrong direction. Another time, Flood saw von Neumann one morning and noticed his eyes were red and swollen. He asked about it, and von Neumann answered simply, "I've been crying." "You didn't ask von Neumann what he was crying about," Flood explains, so the discussion ended there. Incredibly, Flood thinks this was a practical joke, that von Neumann put something in his eyes just so he would be asked about it.

Of course, telling jokes or playing practical jokes is not the same thing as having a sense of humor. Von Neumann's sense of humor was often sarcastic. Once on a train he asked the conductor to send the man with the sandwich tray. The harried conductor said he would tell the sandwich man if he saw him. Von Neumann replied archly, "This train is linear, isn't it?" Much of his humor was like that, witty, but also an insensitive thing to say out loud under the circumstances. Jokes at the expense of Klara and professional colleagues could be cruel. Herman Goldstine recalled that he and von Neumann would attend boring lectures and count the number of times the speaker said something "nebech." They would flash the count by holding up fingers.

There was an element of whimsy in his makeup, too. On one occasion, von Neumann and a colleague spent an evening searching out prime numbers in their personal lives: phone numbers, street addresses, etc. Von Neumann liked hearing and spreading gossip. Of this Ulam claimed oddly that "one often had the feeling that in his memory he was making a collection of human peculiarities as if preparing a statistical study."

Von Neumann was politically conservative, but no knee-jerk reactionary. He would defend Oppenheimer against his attackers. When physicists Ernest O. Lawrence and Luis Alvarez were trying to convince Edward Teller to establish a laboratory at Livermore, von Neumann advised Teller, "Edward, don't join those people—they are too reactionary."

The von Neumanns appreciated luxury. Their china and silver were expensive by the standards of college professors. Von Neumann drove late-model Cadillac convertibles and also had a Studebaker. In the 1940s, when cars first appeared with the gadget that squirts cleaning fluid on the windshield, von Neumann was one of the first to get one, and drove around Princeton showing it off to friends.

The founder of game theory enjoyed games and toys—although such claims may have been overstated in the press in the interest of good copy. The aforementioned *Good Housekeeping* article stated, "Dr. von Neumann loves ingenious children's toys. His friends give him a vast assortment of them on his birthdays, partly as a gag, partly because they know he really adores them. He has been observed unaffectedly scrapping with a five-year-old over who was to have priority in using a new set of interlocking building blocks. He never patronizes children . . ." A birthday present he enjoyed was a toy that, when plucked with a finger, mechanically chirped, "Happy Birthday!"

One of von Neumann's theoretical interests was the question of whether a machine could be built to reproduce itself. He bought a box of Tinker Toys to play with, as a way of addressing this question. He never got very far with a self-reproducing Tinker Toy construction, and ended up giving the set to Oskar Morgenstern's son Karl.

Von Neumann enjoyed food and drink. "He can count anything except calories," quipped the *Good Housekeeping* article. His "dieting" was described as having yogurt and hard-boiled eggs for breakfast and anything he wanted the rest of the day. The article reported von Neumann "likes sweets and rich dishes, preferably with good nourishing sauce, based on cream. He loves Mexican food. When he was stationed at Los Alamos on the bomb project, he would drive 120 miles to dine at a favorite Mexican restaurant." Another time he ordered a "brandy with a hamburger chaser."

Stories paint von Neumann as an occasional heavy drinker. As with so much about him, it is difficult to judge von Neumann by the usual standards. Drinking was part of the Princeton milieu. Raoul Bott spoke of an institute belief in the "therapeutic effects" of alcohol on a tired mathematical mind. J. Robert and Kitty Oppenheimer's house was nicknamed "Bourbon Manor." Liquor flowed freely at the von Neumann household, too, and the host partook of it no less than his guests did. Von Neumann had one of those teetering glass birds that periodically dips its bill in a glass of water. At one party, Johnny decreed that everyone take a drink each time the bird did.

His brother Nicholas, who lived in the von Neumann household at Princeton for several years, told me that Johnny pretended to be drunk in order "to get along" with his peers. Other stories go far beyond conviviality. It was claimed that von Neumann could drink a *quart* of rye whiskey in an hour, and further that he would drive after such a stunt. This may, of course, be exaggeration. Neither written sources nor anyone I talked to suggested that von Neumann's drinking interfered with his work.

Von Neumann was an aggressive and apparently reckless driver. He supposedly totaled a car every year or so. An intersection in Princeton was nicknamed "Von Neumann Corner" for all the auto accidents he had there. Records of accidents and speeding arrests are preserved in his papers. On November 16, 1950, von Neumann had an accident requiring minor body work. On October 23, 1951, he was fined ten dollars for a traffic violation. On May 19, 1953, he was arrested for speeding on the West Side Highway in New York. On July 15, 1953, not two months later, he hit the door of a parked car in Santa Monica.

Though one wonders, there is no evidence that alcohol was a factor in these accidents. Cuthbert Hurd, then director of International Business Machines Corporation's applied science department, told me that von Neumann's problem was not drinking but rather *singing* while driving: he would sway back and forth, turning the steering wheel in time with the music. When von Neumann drove to IBM's offices in Poughkeepsie, New York, he routinely got cited for moving violations in New York City and submitted the tickets to IBM, which had a downtown Manhattan office conveniently near City Hall.

The fascination of John von Neumann derives from his contradictions. He was a mild, charming man who conceived starting a nuclear war and suspected the human race was doomed by its misuse of technology. But one searches long and hard for unambiguous evidence of a darker side to his personal relations. Most who knew him held him in the highest admiration. Historian Laura Fermi (wife of the physicist, Enrico) wrote in *Illustrious Immigrants* that von Neumann "was one of the very few men about whom I have not heard a single critical remark. It is astonishing that so much equanimity and so much intelligence could be concentrated in a man of not extraordinary appearance."

There were many small acts of kindness and generosity. In 1946 von Neumann sent $20 to his former teacher Feher upon learning of a financial reverse. In 1954 he asked the institute to transfer $3,500 allocated to him to the visiting Japanese mathematician Hirotada Anzai.

Von Neumann nursed a minor feud with mathematician Norbert

Wiener. At one lecture of Wiener's, von Neumann sat up front and noisily read the *New York Times*. But Wiener was hardly an enemy. Wiener once tried to get the von Neumanns an invitation to visit China, and described them to Yuk Wing Lee of Tsing Hua University in flattering terms (letter dated May 4, 1937, M.I.T. Archives): "Neumann is one of the two or three top mathematicians in the world, is totally without national or race prejudice, and has an enormously great gift for inspiring younger men and getting them to do research. . . . The Neumanns rather like to hit the high spots socially. You know Princeton life is a bit fast and 'cocktail partyish.' On the other hand, Neumann is not high-hat in any way, and is most accessible to young students."

A flip and rather puzzling comment on von Neumann occurs in physicist Richard Feynman's popular book *Surely You're Joking, Mr. Feynman!* (1985). Feynman says that "Von Neumann gave me an interesting idea: that you don't have to be responsible for the world that you're in. So I have developed a very powerful sense of social irresponsibility as a result of Von Neumann's advice. It's made me a very happy man ever since. But it was Von Neumann who put the seed in that grew into my *active* irresponsibility!"

Von Neumann's second marriage lasted to the end of his life. The arguments continued, and the fault was surely on both sides. "I hope you have forgiven my modest venture in double crossing," one letter of Johnny's says. Another admits: "We both have nasty tempers, but let's quarrel less. I really love you, and, within the limitations of my horrible nature, I *do* want to make you happy—as nearly as possible, as much of the time as possible."

What was the horrible nature? In an interview with journalist Steve J. Heims (in *John Von Neumann and Norbert Wiener: From Mathematics to the Technologies of Life and Death,* 1980), Eugene Wigner asserted that "Johnny believed in having sex, in pleasure, but not in emotional attachment. He was interested in immediate pleasures but had little comprehension of emotions in relationships and mostly saw women in terms of their bodies." Wigner suggested that von Neumann's real love was his mother. Von Neumann's mother, known as "Gittush," was in many respects the center of the family von Neumann had brought to America. Heims further wrote, "Yet some of his colleagues found it disconcerting that upon entering an office where a pretty secretary was working, von Neumann habitually would bend way over, more or less trying to look up her dress."

Whether such claims are overemphasized is hard to say, but one source of domestic friction is beyond dispute. Von Neumann was a workaholic of the most extreme sort. Like Edison, he slept only a few hours a night, and worked most of his waking hours. In *Adventures of a Mathematician,* Ulam noted that von Neumann "may not have been an easy person to live with—in the sense that he did not devote enough time to ordinary family affairs. . . . Von Neumann was so busy . . . he probably could not be a very attentive 'normal' husband. This might account in part for his not-too-smooth home life."

While von Neumann and Morgenstern were working on their game theory treatise in marathon sessions that ran from breakfast to evening parties, Klara became so exasperated that she declared that she didn't want to have anything more to do with game theory unless it "included an elephant." The elephant turns up on page sixty-four of *Theory of Games and Economic Behavior,* hidden in an arcane diagram.

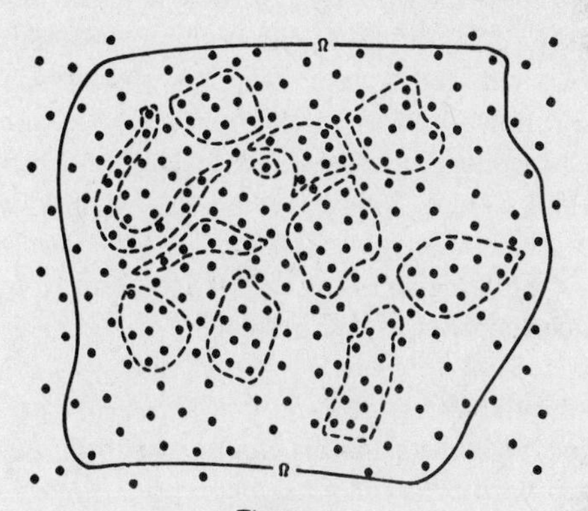

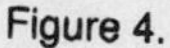

Figure 4.

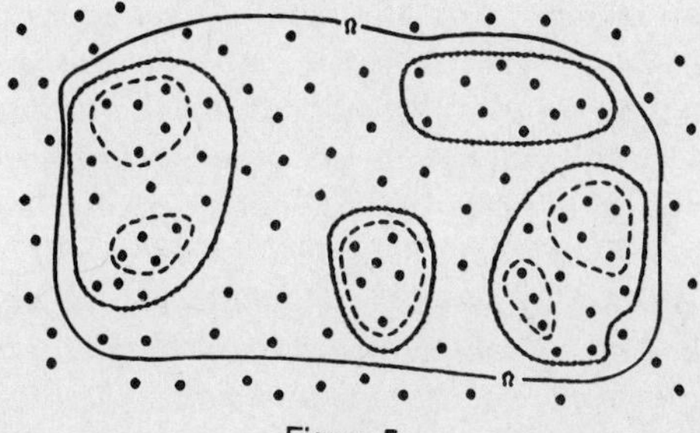

Figure 5.

THE STURM UND DRANG PERIOD

"For Von Neumann the road to success was a many-laned highway with little traffic and no speed limit," swelled his obituary in *Life* magazine. J. Dieudonné's biographical entry in the *Dictionary of Scientific Biography* colorfully calls the fifteen years from 1925 through 1940 von Neumann's *Sturm und Drang* period. He produced original papers in rapid succession in a variety of fields: logic, set theory, group theory, ergodic theory, and operator theory. In a paper published in *Science* shortly after von Neumann's death, Goldstine and Wigner asserted that von Neumann had made significant contributions to every field of mathematics with the exception of topology and number theory.

Throughout his life von Neumann leaned toward applied mathematics, or pure mathematics with recognizable applications. In an essay titled "The Mathematician" (included in James Newman's *The World of Mathematics*), von Neumann describes his personal concept of mathematics. It shows him to be thoughtful and original concerning the philosophical underpinnings of his discipline. There is a curious duality in this short piece. One word he keeps using is *aesthetical.* He defends in Whistlerian terms mathematics for mathematics' sake—consciously posing analogies to the visual arts. Von Neumann lists the qualities of a good mathematical proof:

> *One also expects "elegance" in its "architectural," structural make-up. Ease in stating the problem, great difficulty in getting hold of it and in all attempts at approaching it, then again some very surprising twist by which the approach, or some part of the approach, becomes easy, etc. Also, if the deductions are lengthy or complicated, there should be some simple general principle involved, which "explains" the complications and detours, reduces the apparent arbitrariness to a few simple guiding motivations, etc. These criteria are clearly those of any creative art, and the existence of some underlying empirical, worldly motif in the background—overgrown by aestheticizing developments and followed to a multitude of labyrinthine variants—all this is much more akin to the atmosphere of art pure and simple than to that of the empirical sciences.*

At the same time he insists that the best mathematics is usually inspired by practical problems. This can be read as a defense of game theory (among other things) to those fellow mathematicians who deprecated it as an applied field. Von Neumann warns that "pure" mathematics

> *. . . becomes more and more purely aestheticizing, more and more purely* l'art pour l'art. *This need not be bad, if the field is surrounded by correlated subjects, which still have closer empirical connections, or if the discipline is under the influence of men with an exceptionally well-developed taste. But there is a grave danger that the subject will develop along the line of least resistance, that the stream, so far from its source, will separate into a multitude of insignificant branches, and that the discipline will become a disorganized mass of details and complexities. In other words, at a great distance from its empirical source, or after much "abstract" inbreeding, a mathematical subject is in danger of degeneration. At the inception the style is usually classical; when it shows signs of becoming baroque, then the danger signal is up. It would be easy to give examples, to trace specific evolutions into the baroque and the very high baroque . . .*

Unfortunately, it is impossible to give a sense of von Neumann's mathematical aesthetics without delving into heavy mathematics. All of this work is forbidding to the nonmathematician, and most does not bear on the current discussion. It would be remiss not to mention, however, some of his accomplishments. The reputation von Neumann built in his early years paved the way for the acceptance of game theory.

Set theory was one of his first interests. At the age of twenty, von Neumann devised the formal definition of ordinal numbers that is used today: an ordinal number is the set of all smaller ordinal numbers.

At Göttingen, David Hilbert communicated to von Neumann his enthusiasm for physics and the axiomatic approach to mathematics. Hilbert admired Euclid's *Elements of Geometry,* the geometry book of antiquity (c. 300 B.C.). Euclid's statements about geometry are neither his personal opinions nor the empirical result of careful measurement of geometrical drawings. They are *theorems,* results proved in a chain of logical reasoning. Euclid's work introduced the notion of mathematical proof in a concise, appealing format.

The proofs in today's mathematical journals are more rigorous than Euclid's, but they share the same basic plan. It is impossible to prove anything without some starting point. There must be some accepted body of fact so unquestioned that it may serve as a foundation for proofs of more controversial claims. Euclid called these accepted facts *axioms.*

Today's mathematicians take a broader view of axioms than Euclid did. To Euclid an axiom was a statement so obvious it *had* to be true. Contemporary workers often adopt axioms that do not apply to the observable world to see what may be proved from them. In either case, a proof is most appealing when it requires as few axioms as possible. All the geometry in Euclid's *Elements* (and much more) is derived from just five axioms. The number of axioms in other fields is usually comparably small.

Over the centuries, Euclid's axiomatic approach has appealed to thinkers in many fields. Between 1910 and 1913, British mathematicians Bertrand Russell (whose life intertwines with von Neumann's in minor but striking ways) and Alfred North Whitehead published their multivolume *Principia Mathematica.* This ambitious book was an attempt to axiomize all of mathematics. It showed that much of mathematics could be derived from a few axioms of logic.

We all tend to suppose that mathematics has a strictly logical foundation. That's what makes it mathematics and not a physical science. So there was nothing extraordinary about the idea. Russell and Whitehead's book broke new ground with its scope. They developed the idea further than anyone had previously, surmounting several unexpected obstacles along the way.

But did they succeed? Hilbert realized that they did not. Russell and Whitehead had still not proved that every mathematical truth could be derived using the scheme in *Principia Mathematica,* or even that only true statements were derivable. Hilbert challenged his brilliant protégés to prove that mathematics could be axiomized.

Von Neumann took up the challenge—all for naught. The whole of mathematics *can't* be axiomized, and Kurt Gödel rocked the world of mathematics with a demonstration of that in 1931.

Von Neumann told a story about this ultimately doomed work on Hilbert's suggestion. He came to a sticking point in his proof (he was trying to prove that a system of mathematics similar to *Principia*

Mathematica is self-consistent, which is now known to be unprovable). That night von Neumann had a dream telling him how to continue with the proof. He woke up, got pencil and paper, and carried the proof several stages further. Then he came to another dead end. He went back to sleep. The next day he was unable to make any headway. That night he had another brilliant idea in a dream. He got up again and added another step to the proof. The proof, of course, was still incomplete. Von Neumann concluded, "How lucky mathematics is that I didn't dream the third night!"

What many mathematicians deem von Neumann's greatest, most original achievements are difficult even to describe to the layperson. I will just list them: the theory of "rings of operators," now known as von Neumann algebras, the proof of the quasi-ergodic hypothesis (1932), and work in lattice theory (1935–37).

Beginning in 1927, von Neumann applied the axiomatic approach to the new discoveries in physics. He realized that the state of a quantum mechanical system can be treated as a vector in Hilbert space. This was quintessential von Neumann. Some of his colleagues believe that no one else would have made this discovery for years. Beyond its enormous technical importance, von Neumann's treatment of quantum theory helped emphasize the irreconcilable strangeness of the theory. Von Neumann's work was influential in subsequent "philosophical" interpretations of the theory. As von Neumann saw it, a physical observation involves an observer, a measuring instrument, and that which is being observed. Von Neumann asserted that the distinction between the observer and the measuring instrument is arbitrary.

This work explored the role of the mind in the world, which was to recur in different contexts. In the late 1920s, von Neumann was already working on game theory, and he would spend much of the last years of his life considering how a "mind" of sorts might be embodied in the relays and circuits of a computer. It was these two ideas, game theory and the computer, to which Jacob Bronowski referred in *The Ascent of Man* when he called von Neumann "the cleverest man I ever knew, without exception. And he was a genius, in the sense that a genius is a man who has *two* great ideas."

THE BEST BRAIN IN THE WORLD

Von Neumann's brilliant accomplishments can only hint at the academic and public perception of him. To many of his associates at Princeton, the Pentagon, the RAND Corporation, and elsewhere, von Neumann was a living legend whose reputation was spread by innumerable anecdotes, some of which appeared in the popular press. A magazine article credited him with "the best brain in the world." A Princeton joke had it that von Neumann was not human but a demigod who had made a detailed study of humans and could imitate them perfectly. To get the full import of this pleasantry, remember that von Neumann was not the only genius in town. This was the Princeton of Einstein and Gödel.

Separating fact from legend with von Neumann is a maddening task. He was a genius, a practical joker, and a raconteur, all of which tend to produce stories that sound too pat, too anecdote-like, to be true. I ran the best-known anecdotes by several surviving associates. All had heard of the stories; few could supply any specifics—except to tell new anecdotes. In general the people who actually knew von Neumann were less skeptical of the stories than I was. So all right: whether streamlined in the retelling or not, these stories give an idea of how von Neumann was perceived by others and possibly himself.

Von Neumann's memory was a big part of his legend. It is difficult to demonstrate genius at a faculty tea or a cocktail party; memory is something else. True hypermnesiacs (persons gifted with "photographic" memory) are rare. By no means all are better off for their ability. The memory of the famous patient "S." of Russian psychologist A. R. Luria led to mythic tragedy. S. grew unable to distinguish present experiences from his too-vivid recollections of the past, and spent his last years in an insane asylum. Fortunately, von Neumann's memory was more selective. Klara claimed that her husband wouldn't remember what he had for lunch but could recall every word of a book read fifteen years ago.

Herman Goldstine confirms this seemingly hyperbolic statement in his book *The Computer from Pascal to von Neumann* (1972): "As far as I could tell, von Neumann was able on once reading a book or article to quote it back verbatim; moreover he could do it years later without hesitation. He could also translate it at no diminution in speed from its original language into English. On one occasion I tested his ability by asking him to tell me how the *Tale of Two Cities* started. Whereupon, without any pause, he immediately began to recite the first chapter and continued until asked to stop after about ten or fifteen minutes."

Over the years von Neumann consumed most of the well-known encyclopedic histories, from Gibbon's *Decline and Fall of the Roman Empire* to the *Cambridge Ancient* and *Medieval History* set. "He is a major expert on all the royal family trees in Europe," said an unnamed friend quoted in *Life* magazine. "He can tell you who fell in love with whom, and why, what obscure cousin this or that czar married, how many illegitimate children he had and so on." On a trip through the American South, von Neumann astounded Ulam with his ability to recount the most minute history of the battlefield sites they passed. It is worth emphasizing that von Neumann's military and political views cannot be dismissed out of hand as those of a physical scientist speaking out of his field. Von Neumann's knowledge of history seems to have been at least commensurate with that of professionals in the field.

One of the von Neumann anecdotes claims just that. A famous expert on Byzantine history attended a party at the von Neumanns' house in Princeton. Von Neumann and the expert got into a historical discussion and differed over a date. They pulled out a book to check it, and von Neumann was right. Several weeks later, the history professor was invited to the von Neumanns' again. The professor called Klara and said, "I'll come if Johnny promises not to discuss Byzantine history. Everybody thinks I am the world's greatest expert in it, and I want them to keep on thinking that."

Even von Neumann's memory had its limits. Near the end of his life, von Neumann complained that pure mathematics had burgeoned to the point where it was not possible for any one person to be familiar with more than one quarter of the field. The implication was that he was speaking from experience; *he* was that one person who knew one quarter of mathematics. He comes off more mortal in a few tales of professorial absentmindedness. Klara told of a time when Johnny left his home for an appointment in New York. Sometime later he called her from New Brunswick, New Jersey, to ask, "Why am I going to New York?"

Von Neumann was a calculating prodigy as well. He could divide two eight-digit numbers in his head with little effort. Cuthbert Hurd of IBM told me of von Neumann's uncanny ability to create and revise computer programs (as long as fifty lines of *assembly-language* code!) in his head. He committed to memory a wide selection of physical and mathematical constants. It's easy to imagine how this, combined with his computational abilities, allowed him to produce mental feats that border on the incredible.

An element common to many of the von Neumann anecdotes is the portrayal of him as not merely brilliant but as capable of solving in a flash problems that other bright and educated people cannot solve with long drudgery. His ability to solve other people's problems recommended him to the industrial and military establishments, and must have predisposed him toward the direction his career took.

Goldstine told Heims, "If he would press on a problem, and he couldn't push it through, it would be put aside. It might be two years later when suddenly you'd get a phone call—it might be from God knows where at two in the morning, and there was von Neumann on the wire: 'I now understand how to do such and such.' "

In 1954 a physicist working on the ICBM project for a West Coast aerospace company approached von Neumann with a detailed plan for a certain part of the project. This was a weighty book hundreds of pages long, the product of eight months' work. Von Neumann started leafing through the book. Part of the way through, he turned to the last page and started skimming backward from the end. He wrote down several figures on a pad. Finally he said, "It won't work." The physicist was disappointed, but refused to be deterred. He spent two more grueling months on the project, then decided it wouldn't work after all.

A rare mention of von Neumann's fallibility is as the butt of a practical joke. A typical version of this oft-told but variable tale goes: once a young scientist at the Aberdeen Proving Ground worked out the details of a mathematical problem beforehand and came up to von Neumann at a party. The scientist presented the problem as one that had been stumping him. Von Neumann gazed into the middle distance and began to calculate. Just as he was about to arrive at each intermediate result, the scientist interrupted, "It comes to this, doesn't it?" Of course he was right each time. Finally the young scientist beat von Neumann to the answer. Von Neumann was shaken until he found out it was a setup.

One final story involves a young Raoul Bott (now a prominent Harvard mathematician). In a 1984 speech, Bott told how he became an institute legend by asking von Neumann (in a state of mutual inebriation) what it felt like to be a big shot. Bott's recollection ends on a note of melancholy not inappropriate to a man who would become one of the century's great pessimists.

Some aspects of the cocktail party are blurry in my memory now —doubtless due to the long time span that has elapsed—but I remember distinctly that eventually a small number of us remained playing marbles on Deane [Montgomery]'s carpet. Von Neumann was one of the party, and somehow in the course of the game I came to ask him what it felt like to be a "great mathematician." In his characteristic quiet, thoughtful way, it behooved him to take my question seriously—although a moment earlier he had given us a sampling of his seemingly unlimited store of raunchy stories. Essentially what he said was that, in all honesty, he had known only one "great mathematician"—David Hilbert. And that as far as he was concerned, what with being a Wunderkind, he never really felt that he had lived up to what had been expected of him.

3

GAME THEORY

The idea of a game mirroring the conflicts of the world is an old one. In the *Mabinogion,* a collection of Welsh folktales (eleventh to thirteenth centuries), one story has two warring kings playing chess while their armies battle nearby. Each time one king captures a piece, a messenger arrives to inform the other that he has lost a crucial man or division. Finally one king checkmates. A bloody messenger staggers in and tells the loser, "The army is in flight. You have lost the kingdom."

This fiction refers to the frankly military origins of chess. The Chinese game of go, the Hindu chaturanga, and many other games are battle simulations, too. Those who see games as simulations of war may see war as a kind of game, too. The classic instance of this was Prussia's century-long infatuation with *Kriegspiel.*

KRIEGSPIEL

Devised as an educational game for military schools in the eighteenth century, Kriegspiel was originally played on a board consisting of a map of the French-Belgian frontier divided into a grid of 3,600 squares. Game pieces advanced and retreated across the board like armies.

The original Kriegspiel spawned many imitations and was ultimately supplanted by a version that became popular among Prussian army officers. This used real military maps in lieu of a game board. In 1824 the chief of the German general staff said of Kriegspiel, "It is not a game at all! It's training for war!"

So began a national obsession that defies belief today. The Prussian high command was so taken with this game that it issued sets to every army regiment. Standing orders compelled every military man to play it. The Kaiser appeared at Kriegspiel tournaments in full military regalia. Inspired by overtly militaristic chess sets then in vogue (pieces were sculpted as German marshals, colonels, privates, etc.), craftsmen produced Kriegspiel pieces of obsessive detail. A pale remnant of these *Zinnfiguren* ("tin figures") survives today as toy soldiers. Layer after layer of complexity accreted around the game as its devoted players sought ever greater "realism." The rule book, originally sixty pages, grew thicker with each edition. Contingencies of play that were once decided by chance or an umpire were referred to data tables drawn from actual combat.

Claims that the game was behind Prussia's military victories stimulated interest internationally. Prussia's Kriegspiel dry runs of war with Austria supposedly led to a strategy that proved decisive in the Six Weeks' War of 1866. After that, the Austrian Army took no chances and began playing Kriegspiel. France's defeat in the Franco-Prussian War (1870)—allegedly another Kriegspiel victory for Prussia—spawned a Kriegspiel craze there.

Kriegspiel came to the United States after the Civil War. One American army officer complained that the game "cannot be readily and intelligently used by anyone who is not a mathematician, and it requires, in order to be able to use it readily, an amount of special instruction, study, and practice about equivalent to that necessary to acquire a speaking knowledge of a foreign language." Nonetheless, it eventually became popular in the Navy and at the Naval War College in Newport, Rhode Island.

Japan's victory in the Russo-Japanese War (1905) was the last credited to a nation's playing of Kriegspiel. It became apparent that strategies honed in the game did not always work in battle. Germany's defeat in World War I was a death knell for the game—except, ironically, in Germany itself, where postwar commanders fought each other with tin replicas of the regiments denied them by the Treaty of Versailles.

In Budapest, the young John von Neumann played an improvised Kriegspiel with his brothers. They sketched out castles, highways, and coastlines on graph paper, then advanced and retreated "armies" according to rules. During World War I, Johnny obtained maps of the fronts and followed reports of real advances and retreats. Today Kriegspiel is usually played with three chessboards, visible only to an umpire. In this form it was a popular lunch-hour pastime at the RAND Corporation, and von Neumann played the game on visits there.

To some critics, game theory is the twentieth century's Kriegspiel, a mirror in which military strategists see reflected their own preconceptions. The comparison is revealing even while being unfair. Game theory did become a kind of strategic oracle, particularly in the two decades after Hiroshima. The problem is one common to oracles, namely that game theory's answers can depend on exactly how you phrase the questions.

Why a theory of games? It is a cliché of scientific biography to find reasons in a scientist's personality for choosing a subject. But the question is fair enough. Though the scientist or mathematician is a discoverer rather than a creator, there is a literal universe of avenues to explore. Why one and not another?

Meaningful answers are harder to come by than historians of science like to admit. When the question has been put to living scientists, they are often at a loss for words. Many have noted von Neumann's fascination with play, his collection of children's toys, his sometimes childish humor. In this he was not atypical among scientists. Jacob Bronowski wrote (1973), "You must see that in a sense all science, all human thought, is a form of play. Abstract thought is the neoteny[1] of the intellect, by which man is able to carry out activities which have no immediate goal (other animals play only while young) in order to prepare himself for long-term strategies and plans."

Game theory is not about "playing" as usually understood. It is about conflict among rational but distrusting beings. Von Neumann escaped revolution and terrorism in Hungary and later the rise of Nazism. His relationship with Klara was one of repeated conflict. In his letters to his wife Johnny talks of double-crossing, reprisals, and boundless distrust. That's part of what game theory is about.

Game theory was the brainchild of a cynic. Some commentators have suggested that von Neumann's personal cynicism influenced the theory. It is conceivable that von Neumann's personality led him to explore game theory rather than something else. It is wrong to think that von Neumann concocted game theory as a "scientific" basis for his personal beliefs or politics. Game theory is a rigorously mathematical study which evolves naturally from a reasonable way of looking at conflict. Von Neumann would not have pursued game theory had his mathematical intuition not told him that it was a field ripe for development. Some of the mathematics of game theory are closely related to that von Neumann used in treating quantum physics.

1. Retention of immature traits in adulthood. Bronowski alludes to the fact that nonhuman animals play and experiment in their youth, then lock into a successful pattern of behavior (compare the playful kitten with the contented old cat).

The nominal inspiration for game theory was poker, a game von Neumann played occasionally and not especially well. (A 1955 *Newsweek* article appraised him as "only a fair-to-middling winner" at the game.) In poker, you have to consider what the other players are thinking. This distinguishes game theory from the theory of probability, which also applies to many games. Consider a poker player who naively tries to use probability theory alone to guide his play. The player computes the probability that his hand is better than the other players' hands, and wagers in direct proportion to the strength of the hand. After many hands, the other players will realize that (say) his willingness to sink twelve dollars in the pot means he has at least three of a kind. As poker players know, that kind of predictability is bad (a "poker face" betrays nothing).

Good poker players do not simply play the odds. They take into account the conclusions other players will draw from their actions, and sometimes try to deceive the other players. It was von Neumann's genius to see that this devious way of playing was both rational and amenable to rigorous analysis.

Not everyone agreed that game theory was the most fruitful outlet for von Neumann's ample talents. Paul Halmos, von Neumann's assistant at Princeton, told me, "As far as I was concerned, he was just wasting his time on 'that game stuff.' I know full well that a large part of the world doesn't agree with the opinions I held then, and I am not sure whether I myself agree with them now, but . . . I never learned the subject and never learned to like it."

WHO WAS FIRST?

Von Neumann cannot be given undivided credit for the invention of game theory. Beginning in 1921, seven years before von Neumann's first paper, French mathematician Émile Borel published several papers on *"la théorie du jeu."* The parallels between these papers and von Neumann's work are strong. Borel used poker as an example and took up the problem of bluffing, just as von Neumann would. Borel appreciated the potential economic and military applications of game theory. Indeed, Borel warned against overly simplistic applications of game theory to warfare. He was not talking off the top of his head. Borel, who had held public office, became minister of the French Navy in 1925. Most important, Borel posed the basic questions of game theory: for what games is there a best strategy, and how does one find such a strategy?

Borel did not develop these issues very far. Like so many creative individuals, von Neumann was jealous of prior claims to "his" innovation. His 1928 paper and the 1944 book make but scant mention of Borel, that in footnotes. Lest there be any doubt, Ulam said that one of Borel's papers had indeed inspired von Neumann's work.

Von Neumann's slighting treatment of Borel long contributed to an underappreciation of the latter's work. In 1953 von Neumann was reportedly furious to learn that Borel's papers were being translated into English. The translator, mathematician L. J. Savage, told Steve Heims: "He phoned me from someplace like Los Alamos, very angry. He wrote a criticism of these papers in English. The criticism was not angry. It was characteristic of him that the criticism was written with good manners."

All this granted, the seminal paper of game theory is without doubt von Neumann's 1928 article, *"Zur Theorie der Gesellschaftspiele"* ("Theory of Parlor Games"). In this he proved (as Borel had not) the famous "minimax theorem." This important result immediately gave the field mathematical respectability.

THEORY OF GAMES AND ECONOMIC BEHAVIOR

Von Neumann wanted game theory to reach a larger audience than mathematicians. He felt the developing field would be of most use to economists. He teamed with an Austrian economist then at Princeton, Oskar Morgenstern, to develop his theory.

Von Neumann and Morgenstern's *Theory of Games and Economic Behavior* is one of the most influential and least-read books of the twentieth century. Princeton University Press admitted as much in an ad it ran in *American Scientist* to commemorate the fifth year of anemic sales. "Great books often take a while to achieve recognition. . . . Then, later, when the world learns about them, their influence far surpasses their readership." The book had still not quite sold 4,000 copies in five years, a fact that would normally be hard to square with the contention that the book had taken the field of economics by storm. Most economists still hadn't read it (and never would); it wasn't even in the libraries of many schools of economics. The ad noted "a few copies bought by professional gamblers."

Theory of Games and Economic Behavior is a difficult book. Today, the reader's enthusiasm for plowing through all 641 formula-filled pages is dampened by the fact that von Neumann and Morgenstern got sidetracked in their treatment of games of more than two persons. Their approach, while not wrong, no longer seems the most useful or illuminating one.

If nothing else, the book is ambitious. Von Neumann and Morgenstern dreamed of doing for economics what von Neumann did for quantum physics and could not do for mathematics itself: putting it on an axiomatic basis. The authors stated: "We hope to establish satisfactorily . . . that the typical problems of economic behavior become strictly identical with the mathematical notions of suitable games of strategy."

Theory of Games and Economic Behavior thus presents itself as a pioneering work of economics. The book's introduction is almost apologetic about investigating recreational games. The games are presented as potential models for economic interactions. (Military applications, which were to become so important to von Neumann's followers, are not mentioned.)

The tone is iconoclastic. Von Neumann and Morgenstern insist that economists must go back to square one. They deride the then-current state of mathematical economics, comparing it with the state of physics before Kepler and Newton. They chide those who advocate economic reform based on presently unconfirmable theories. One gathers the authors were thinking of Marxism, among other theories.

The authors speculate that a future exact science of economics will require its own, yet-unimagined mathematics. They suggest that calculus, which is ultimately derived from the physics of falling and orbiting bodies, is presently overemphasized in mathematics.

Fortunately for our purposes, the essential kernel of game theory is easy to grasp, even for those with little background in—or tolerance for—mathematics. Game theory is founded on a very simple but powerful way of schematizing conflict, and this method can be illustrated by a few familiar childhood games.

CAKE DIVISION

Most people have heard of the reputed best way to let two bratty children split a piece of cake. No matter how carefully a parent divides it, one child (or both!) feels he has been slighted with the smaller piece. The solution is to let one child divide the cake and let the other choose which piece he wants. Greed ensures fair division. The first child can't object that the cake was divided unevenly because he did it himself. The second child can't complain since he has his choice of pieces.

This homely example is not only a game in von Neumann's sense, but it is also about the simplest illustration of the "minimax" principle upon which game theory is based.

The cake problem is a conflict of interests. Both children want the same thing—as much of the cake as possible. The ultimate division of the cake depends both on how one child cuts the cake and which piece the other child chooses. It is important that each child anticipates what the other will do. This is what makes the situation a game in von Neumann's sense.

Game theory searches for *solutions*—rational outcomes—of games. Dividing the cake evenly is the best strategy for the first child, since he anticipates that the other child's strategy will be to take the biggest piece. Equal division of the cake is therefore the solution to this game. This solution does not depend on a child's generosity or sense of fair play. It is enforced by both children's self-interest. Game theory seeks solutions of precisely this sort.

RATIONAL PLAYERS

With this example in mind, let's go back and examine some of the ideas we have introduced. There are many ways of playing games. You can play just for fun with no thought of winning or losing. You may play recklessly, in the hope of lucking out and winning. You may play on the assumption that your opponent is foolish and attempt to exploit that foolishness. In a game of ticktacktoe with a child, you might even play to lose. This is all well and fine. It is not what game theory is about.

Game theory is about *perfectly logical players interested only in winning.* When you credit your opponent(s) with both rationality and a desire to win, and play so as to encourage the best outcome for yourself, then the game is open to the analysis of game theory.

Perfect rationality, like perfect *anything,* is a fiction. There's no such thing as a perfectly straight line. This didn't stop Euclid from developing a useful system of geometry. So it was with von Neumann and his perfectly rational players. You can think of the players of game theory as being something like the perfect logicians you hear about in logic puzzles, or even as being computer programs rather than human beings. The players are assumed to have perfect understanding of the rules and perfect memory of past moves. At all points in the game they are aware of all possible logical ramifications of their moves and their opponents' moves.

This can be a stringent requirement. Perfectly rational players would never miss a jump in checkers or "fall into a trap" in chess. *All* legal sequences of moves are implicit in the rules of these games, and a perfectly logical player gives due consideration to *every* possibility.

But as anyone who plays chess or checkers knows, traps and missed moves—trying to get your opponent to fall for them, trying to recover when you fall for them—are pretty much what the games are all about. What would a game between two perfectly rational players be like?

You probably already know what happens when ticktacktoe is played "rationally." It ends in a draw—it has to unless someone makes a mistake. Because ticktacktoe is so simple that it is possible to learn to play it perfectly, the game soon loses its appeal.

Von Neumann showed, however, that many other games are like ticktacktoe in this sense. Chess is not a game, von Neumann told Bronowski. He meant that there is a "correct" way(s) to play the game—although no one presently knows what it is—and that the game would therefore be trivial, in much the same sense as ticktacktoe, to players aware of the "correct" strategy.

GAMES AS TREES

The gist of von Neumann's demonstration of this fact is marvelously simple. It applies not only to chess but to any game where no information is kept from the players, where "all the cards are on the table."

Most familiar games take place as a sequence of moves by the players. In ticktacktoe, chess, or checkers, the grid or board is always visible to both players. No moves are taken in secret. For any such game, you can draw a diagram of all the possible courses of play. I'll use ticktacktoe as an example because it's fairly simple, but the same thing could be done, in principle, for chess, checkers, or any such game. Ticktacktoe starts with the first player ("X") putting a mark in any of nine cells. There are consequently nine possible first moves. The nine choices open to Player X on the first move can be diagrammed as nine lines radiating up from a point. The point represents the move, the moment of decision, and the lines represent the possible choices.

Next it's Player O's move. There are eight cells still open—which eight depending on where the X is. So draw eight secondary branches at the top of each of the nine primary branches. That leaves seven open cells for X on his second move. As the diagram of possible moves is continued upward, it branches like a very bushy tree.

As you continue the process, you will eventually diagram moves that put three markers in a row. That's a win for the player who moves. It's also the termination of that particular branch in the diagram, for the game ends when someone gets three in a row. Mark that point (call it a "leaf" of the diagram) as a win for X or O as the case may be.

Other branches of the diagram will terminate in a tie. Mark them as ties. Obviously, the game of ticktacktoe cannot go on forever. Nine moves is the maximum. So eventually, you will have a *complete* diagram of the game of ticktacktoe. Every possible ticktacktoe game—every game that ever has been played or ever will be played—must appear in the diagram as a branch starting at the "root" (X's first move) and continuing up to a "leaf" marked as a win for X, a win for O, or a tie. The longest complete branches/games are nine moves long. The shortest are five moves (this is the minimum for a win by the first player).

So much for the tree; now for the pruning shears. By process of elimination, you can figure out how to play ticktacktoe "rationally" from the diagram. The diagram contains all legal sequences of play, even those with *stupid* moves such as when someone overlooks a chance to get three in a row. All you have to do is take pruning shears to the tree and trim off all the stupid moves. What's left will be the smart moves—the rational way to play.

A small portion of the diagram looks like this:

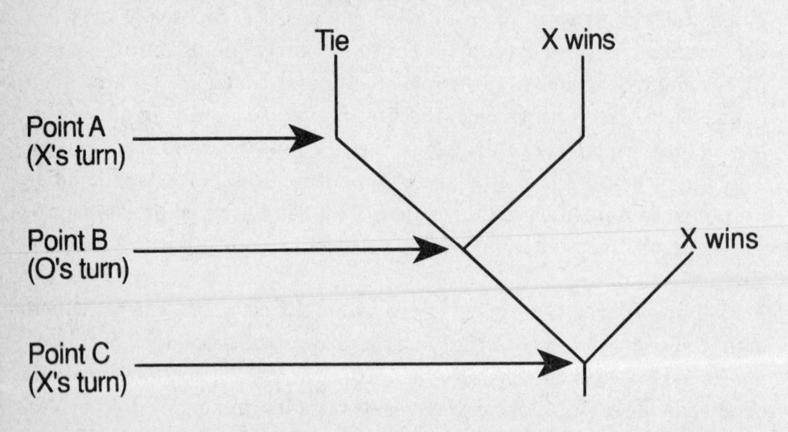

Go through the diagram and carefully backtrack from every leaf. Each leaf is someone's last move, a move that creates a victory or a tie. For instance, at Point A, it is X's move, and there is only one empty cell. X has no choice but to fill it in and create a tie.

Now look at Point B, a move earlier in the game. It is O's turn, and he has two choices. Putting an O in one of the two open cells leads to the aforementioned Point A and a sure tie. Putting an O in the other cell, however, leads to a win for X. A rational O player prefers a tie to an X victory. Consequently, the right branch leading upward from Point B can never occur in rational play. Snip this branch from the diagram. Once the play gets to Point B, a tie is a foregone conclusion.

But look: X could have won earlier, at Point C. A rational X would have chosen an immediate win at Point C. So actually, we can snip off the entire left branch of the diagram.

Keep pruning the tree down to the root, and you will discover that ties are the only possible outcomes of rational play. (There is more than one rational way of playing, though.) The second player can and will veto any attempt at an X victory, and vice-versa.

What can be done for ticktacktoe could be done for almost any two-person game with no hidden information. The main restriction is that the game must be *finite*. It can't go on forever, and the number of distinct choices at any move must also be finite. Otherwise, there are no "leaves" (last moves) to work back from.

Human beings being mortal, no recreational game is intended to go on forever. Rules of more challenging games rarely state a maximum number of moves explicitly, though. Chess normally ends in checkmate. There are many cases where pieces can be moved endlessly without creating a checkmate. Should captures reduce the board to just the two kings, neither will be able to checkmate the other. "Tie rules" bring such games to a halt. A common rule declares the game a tie when a sequence of moves repeats exactly three times. Another, more stringent, rule is that if no pawn is moved and no higher-ranking pieces are captured in forty moves, the game is a tie.

Consequently, von Neumann and Morgenstern pointed out, there is a numerical limit on the number of moves in a game of chess under a given tie rule. (The limiting number is probably around 5,000 moves with typical rules—far more than any game of chess ever played!) The actual value of the limit is not important to the proof, just that it exists and is finite. Given that a game of chess can run only so many moves, and that the number of choices at each move is finite, it follows that the number of different courses of play is itself a finite number. You could make a diagram of all legal courses of play, and prune it to reveal the rational way to play chess.

It recalls the old joke about the chess game where White makes his first move and then Black says, "I resign." Chess between perfectly rational players would be as trivial as that. It is only because we do not know the correct strategy for chess that it still challenges players. It is one thing to prove that a best strategy exists, but it is quite another to do all the calculation and produce the strategy. It is unknown whether a rational game of chess would end in a victory (presumably for White, who moves first) or a tie.

GAMES AS TABLES

There is another way of looking at games, one far more useful in game theory. A game is equivalent to a table of possible outcomes.

As we have shown, the number of possible games of chess is astronomically large but finite nevertheless. It follows that the number of chess strategies is finite, too. I have already used the word "strategy" several times; now is the time to define it. In game theory, strategy is an important idea, and it has a more precise meaning than it usually does. When chess players talk of a strategy, they mean something like "open with the king's Indian Defense and play aggressively." In game theory, a strategy is a much more specific plan. It is a *complete* description of a particular way to play a game, no matter what the other player(s) does and no matter how long the game lasts. A strategy must prescribe actions so thoroughly that you never have to make a decision in following it.

An example of a true strategy for first player in ticktacktoe would be:

> Put X in the center square. O can respond two ways:
>
> **1** If O goes in a noncorner square, put X in a corner cell adjacent to the O. This gives you two-in-a-row. If O fails to block on the next move, make three-in-a-row for a *win.* If O blocks, put X in the empty corner cell that is not adjacent to the first (noncorner) O. This gives you two-in-a-row two ways. No matter what O does on the next move, you can make three-in-a-row after that and *win.*
>
> **2** If instead O's first move is a corner cell, put X in one of the adjacent noncorner cells. This gives you two-in-a-row. If O fails to block on the next move, make three-in-a-row for a *win.* If O blocks, put X in the corner cell that is adjacent to the second O and on the same side of the grid as the first O. This gives you two-in-a-row. If O fails to block on the next move, make three-in-a-row for a *win.* If O blocks, put X in the empty cell adjacent to the third O. This gives you two-in-a-row. If O fails to block on the next move, make three-in-a-row for a *win.* If O blocks, fill in the remaining cell for a *tie.*

This shows how complicated a strategy can be, even for a very simple game. A true strategy for chess would be so huge that it could never be written down. There is not enough paper and ink on earth to list all the possibilities; there is not enough computer memory to loop through them all. This is one reason why computers still aren't unbeatable at chess.

Overwhelming as this practical difficulty is, it didn't bother von Neumann, and it needn't bother us. In fact, since we're fantasizing, we might as well go a step further. Not only could a perfectly rational being conceive of a strategy in full detail; he could—given no limits on memory or computing power whatsoever—anticipate every possible chess strategy and decide in advance what to do even before moving the first piece.

Suppose you had a numbered list of all possible strategies for chess. Then your choice of strategy is tantamount to selecting a number from 1 to *n*, where *n* is the (very, very large) number of possible strategies. Your opponent could choose a strategy from his list of possibilities (from 1 to *m*, say). Once these two strategies were chosen, the resulting game would be completely specified. By applying the two strategies you could move the pieces and bring the game to its preordained conclusion. Openings, captures, "surprise moves," and endgame would all be implicit in the choice of strategies.

To take this pipe dream to its conclusion, we can imagine that, given enough time, you could play out every possible pairing of strategies to see the outcome. The results could be tabulated in a rectangular table. The real table would span the galaxies, so we'll print an abbreviated version here!

		Black's Strategies				
		1	**2**	**3**	**. . .**	***m***
	1	White checkmate in 37 moves	Draw in 102 moves	Black resigns in 63 moves		Black checkmate in 42 moves
White's Strategies	**2**	White checkmate in 45 moves	Black checkmate in 17 moves	White checkmate in 54 moves		White checkmate in 82 moves
	3	White checkmate in 43 moves	White checkmate in 108 moves	Draw in 1,801 moves		Black checkmate in 32 moves
	. . .					
	n	Draw in 204 moves	White checkmate in 77 moves	White checkmate in 24 moves		Draw in 842 moves

Once you had this table, you wouldn't have to bother with the chessboard anymore. A "game" of chess would amount to the two players choosing their strategies simultaneously and looking up the result in the table.[2] To find out who wins, you'd look in the cell at the intersection of the row corresponding to White's strategy and the column of Black's strategy. Should White choose to use strategy number 2 on his list, and Black choose to use his strategy number 3, the inevitable outcome would be a checkmate for White in 54 moves.

This isn't the way real people play real games. To detail every possible contingency beforehand would be the antithesis of the word "play." No matter. This idea of representing games as tables of outcomes turns out to be very useful. Every possible sequence of play for any two-person game can be represented as a cell in a similar type of table. The table must have as many rows as one player has strategies and a column for each of the other player's strategies. A game reduced to a table like this is called the "normalized form" of the game.

The trick is deciding *which* strategy to choose. The table gets all the facts out in the open, but this isn't always enough. The arrangement of outcomes in the table can be capricious. Neither player gets to choose the outcome he wants, only the row or column it appears in. The other player's choice makes an equal contribution.

Look at the imaginary table for chess. Is strategy number 1 a good choice for White? It's tough to say. If Black chooses strategy number 1, it's good because that leads to a win for White. But with other choices for Black, the result can be a draw or a loss for White.

White really wants to determine which strategy Black is going to choose. Then all White has to do is make sure he picks one of his own strategies that will lead to a win when paired with the Black strategy.

Unfortunately, Black wants to do the same thing. Black wants to psych out White, and choose his strategy accordingly for a Black victory. Of course, White knows this, and tries to predict what Black will do based on what he thinks White will do . . .

Borel and von Neumann realized that this sort of deliberation puts games beyond the scope of probability theory. The players would be wrong indeed to assume that their opponent's choice is determined by

2. Why "simultaneously"? Doesn't Black at least get to see White's first move before deciding on his strategy? No: you're failing to appreciate how comprehensive a strategy must be. The first part of a Black strategy would prescribe a Black opening move for *each* of the twenty possible opening moves by White. Not until each of these twenty contingencies is accounted for do you have a strategy in von Neumann's sense.

"chance." Chance has nothing to do with it. The players can be expected to do their very best to deduce the other's choice and prepare for it. A new theory is called for.

ZERO-SUM GAMES

"Zero-sum game" is one of the few terms from game theory that has entered general parlance. It refers to a game where the total winnings or payoffs are fixed. The best example is a game like poker, where players put money in the pot, and someone wins it. No one ever wins a dollar but that someone else has lost it. It is in this restricted but quite diverse category of games that game theory has enjoyed its greatest success. The analogies to economics are obvious. One speaks of a "zero-sum society," meaning that one person's gain is another's loss: "There's no such thing as a free lunch."

Most recreational games are zero-sum. This is true even of games that don't involve money. Whether money is at stake or not, each player prefers some possible outcomes to others. When these preferences are expressed on a numerical scale, they are called *utility*.

Think of utility as the "counters" in a game, or as "points" you try to win. If you play poker for matchsticks, and honestly try to win as many matchsticks as possible, then utility is identical with the quantity of matchsticks.

In a game played for money, money is utility or nearly so. When a game is played just to win, the mere fact of winning confers utility. In a win-or-lose game like ticktacktoe or chess, winning might be assigned a utility of 1 (in arbitrary "points") and losing might be assigned a utility of –1 point. The sum of utilities is still zero, hence it is a zero-sum game.

An important thing to remember about utility is that it corresponds to the actual preferences of the players. In the case of an adult playing a child and wanting to lose, the adult's utilities would be reversed: losing would have a utility of 1 and winning a utility of –1. Thus utility does not necessarily correspond to money, or winning or losing, or any obvious external object.

The simplest true game is a two-person, two-strategy, zero-sum game. The only way a game could be any simpler would be for a player to have just one strategy. But a "choice" of one option is no choice at all. The "game" would really be a one-person game, which is no game at all.

A two-person, two-strategy game can be diagrammed in a two-row by two-column table. If the game is further a zero-sum game, the outcomes can be represented concisely. Fill each of the four cells with a number representing the first player's win. We know that the first player's win is the second player's loss, so both can use the same diagram (the second player's wins are the negatives of the numbers in the table).

MINIMAX AND CAKE

A two-person, zero-sum game is "total war." One player can win only if the other loses. No cooperation is possible. Von Neumann settled on a simple and sensible plan for deciding rational solutions of such games. It is called the minimax principle.

Let's reexamine the cake division problem from the perspective of game theory. The children are playing a zero-sum game. There is only so much cake to begin with, and nothing the children do will change the amount available. More cake for one means that much less cake for the other.

The first child (the "cutter") has a range of strategies—strictly speaking, an unlimited number since he can cut the cake in any of an infinite number of ways. We will not miss much by simplifying the range of choices to just two strategies. One strategy is to split the cake unevenly and the other is to split the cake as evenly as possible.

The second child (the "chooser") also has two strategies. He can choose the bigger piece or the smaller piece. (As a further note of realism, we'll allow that no slicing operation is perfect. So even when the cutter adopts the policy of splitting the cake evenly, one piece *will* be slightly bigger than the other.)

A simple table illustrates the choices. We need put only one child's payoff in the cells of the table. Let's use the cutter's payoff. Obviously, the chooser gets whatever is left. The table looks like this:

		Chooser's strategies	
		Choose bigger piece	Choose smaller piece
Cutter's strategies	Cut cake as evenly as possible	Half the cake minus a crumb	Half the cake plus a crumb
	Make one piece bigger than the other	Small piece	Big piece

We already know what to expect of this game. The cutter will split the cake evenly, or try to as best he can. The chooser will pick the bigger piece. The result is the upper left cell. The cutter will get slightly less than half the cake, since the chooser will take the bigger of two nearly identical pieces.

Why this outcome? If the cutter could have his pick of any of the four possible outcomes, he would want to end up with a big piece (lower right cell). He realizes, however, that this is not realistic. The cutter knows what to expect from the chooser; namely, the worst—as small a piece as possible.

The cutter is empowered only to decide the row of the cake division's outcome. He expects to end up with the *least* amount of cake in that row, for the chooser will act to minimize the cutter's piece. Therefore he acts so as to *maximize* the *minimum* the chooser will leave him.

If the cutter cuts the cake evenly, he knows he will end up with nearly half the cake. If instead the cutter makes one piece much bigger, he knows as certainly that he will end up with a small piece. The real choice, then, is between nearly half the cake and much less than half. The cutter goes for nearly half the cake by electing to split the cake evenly. This amount, the maximum row minimum, is called the "maximin."

"You know that the best you can expect is to avoid the worst," writes Italo Calvino in *If on a Winter's Night a Traveler* (1979). The epigram neatly states the minimax principle. The choice of strategies is a natural outcome. It is not merely a "fair" outcome recommended by game-theoretic arbitration but a true equilibrium enforced by both players' self-interest. A player deviates from his best strategy to his own determent (and to his opponent's benefit because it is a zero-sum game).

The minimax principle helps make sense of more difficult two-person zero-sum games. We have shown that almost any common game is logically equivalent to a *simultaneous* choice of strategies by players. Simultaneous games are thus different from cake division, where the chooser acts *after* the cutter has.

But look: What if the chooser had to go first by announcing his choice (bigger piece or smaller piece) before the cutter picked up the knife? It would make no difference at all. A rational chooser knows the cutter will divide the cake so that the chooser's slice is as small as possible. The chooser in turn wants the cutter to get the smallest piece possible. (Remember, the table above gives the cutter's piece, which is the complement of the chooser's.) The chooser looks for the *minimum* column *maximum* (the "minimax"). It's also the upper left cell. The chooser should go for the bigger piece.

In this game, the upper left cell is the natural outcome, regardless of which player is required to announce his strategy first. Consequently, we feel safe in saying that the upper left cell would be the logical result of a game where players had to decide simultaneously.

The value in the upper left cell is both the maximin (the cutter's best "realistic" outcome) and the minimax (the chooser's best realistic outcome, here expressed as what the cutter would get). You might wonder whether this is a coincidence, or whether it is always so. It is a coincidence, though not an unusual one in a small table. When the maximin and the minimax are identical, that outcome is called a "saddle point." Von Neumann and Morgenstern likened it to the point in the middle of a saddle-shaped mountain pass—at once the maximum elevation reached by a traveler going through the pass and the minimum elevation encountered by a mountain goat traveling the crest of the range.

When a game has a saddle point, the saddle point is the solution of the game, the expected outcome of rational play. Note that a rational solution doesn't necessarily mean that everyone is happy. The cutter ends up getting a crumb or two less than the chooser. He may not think that's fair. For that matter, both players may be disappointed they didn't get a much bigger piece. Neither player gets his first choice of outcome. What's to prevent the players from striking out and doing something irrational?

The answer is greed and distrust. Half the cake minus a crumb is the most the cutter can guarantee for himself without any help from the chooser. It is also the smallest piece the chooser can leave for the cutter by his own efforts. To do any better, a player would need the assistance of his adversary. But the opponent has no reason to help—it's less cake for him. The saddle-point solution of a zero-sum game is self-enforcing. It's something like Chinese finger cuffs. The harder you struggle to do any better, the worse off you are.

MIXED STRATEGIES

Unfortunately, there's a catch. Not all games have saddle points. The trouble is, you can invent a game with any rules you want. Any set of payoffs is conceivable. It is easy to fill a rectangular grid with numbers so that the minimum row maximum does *not* equal the maximum column minimum. Then there is no saddle point.

One of the simplest of all games has no saddle points. "Matching pennies" (which von Neumann and Morgenstern use as an example) is hardly a game at all in the usual sense. Two players simultaneously place a penny on a table—heads up or tails up. When the pennies match (both are heads or both are tails) the first player gets to keep both pennies. He gets back his own penny and wins his partner's penny for a profit of 1 cent. If the pennies don't match, the second player gets both.

The table for matching pennies looks like this:

	Heads	Tails
Heads	1 cent	–1 cent
Tails	–1 cent	1 cent

The minimum of both rows is –1 cent. Therefore the maximum minimum is also –1 cent. The maximum of both columns is 1 cent, so the minimum maximum is 1 cent too. There is a 2-cent difference between the minimax and the maximin.

Von Neumann and Morgenstern likened games to a "tug of war." Each side can prevent the other from gaining too much ground, but there is a middle ground where the rope lurches back and forth. In matching pennies, the first player can guarantee himself his minimax value (–1 cent)—which isn't saying much in this case, because that's the maximum loss in the game. The second player is guaranteed that he can't lose more than a penny. The difference between these two guarantees, 2 cents, is how much is really at stake in the game.

Should you choose heads or tails? Obviously, it all depends on what the other player will do. If you knew what the other player was going to do, you'd know what you want to do—and vice-versa.

As you probably know, the best way to play matching pennies is to play heads and tails *randomly* (with 50 percent probability each). This is called a "mixed strategy" in contrast to the "pure strategies" of playing heads with certainty or tails with certainty. Mixed strategies were nothing new in von Neumann's time. Borel's paper considered such strategies, and of course players of games like matching pennies have long appreciated the desirability of acting randomly. Sometimes matching pennies is used as a "random" way of deciding who gets an advantage in another game, such as which team bats first in baseball.

By fashioning a new, random strategy "from scratch," the players create a self-enforcing equilibrium. Let's make a new diagram for matching pennies that includes the random strategy.

	Heads	**Tails**	**Random**
Heads	1 cent	–1 cent	0
Tails	–1 cent	1 cent	0
Random	0	0	0

Anyone who plays randomly stands equal chances of winning and of losing a penny. (This is true whether the opponent plays a pure strategy or chooses randomly as well.) On the average, then, the payoff to a random player is 0. Fill in the row and column for the random strategies with 0's.

Now there is a saddle point. If the first player had to name his strategy first (definitely heads, definitely tails, or a random choice), knowing that the second would take full advantage of that information, he'd want to choose the strategy with the maximum minimum. The strategies of heads or tails have minimums of –1 cent. The random strategy guarantees an average gain of 0, no matter what the other player does. Thus the random strategy has the maximum minimum.

If the second player had to go first, he would want the minimum maximum. Again this is the random strategy. Game theory suggests the lower right cell as the natural outcome. Both players should choose randomly. Once again we find an equilibrium between the players' opposed interests.

Most five-year-olds already know how to play matching pennies. What do we need game theory for?

The answer is that other games are not quite so simple, and for them game theory can crank out impeccably correct prescriptions that are by no means common sense. The odds in a random strategy do not have to be fifty-fifty. They can and should be adjusted according to the payoffs. Game theory tells how to do this.

Here's a nice little dilemma: "Millionaire Jackpot Matching Pennies." It works just like ordinary matching pennies except that you play against fabulously wealthy opponents only, and if you match on heads, your opponent has to pay you a *million dollars.* Your payoffs are as follows (your opponent's are just the opposite).

	Heads	**Tails**
Heads	$1 million	–1 cent
Tails	–1 cent	1 cent

How should you play this game? Well, the pennies are chicken feed. You're interested in winning that million dollars. The only way that can happen is for you to play heads. So your first impulse is to play heads.

But wait a minute, your opponent would be *crazy* to play heads. He's not going to risk losing a million dollars. His first impulse is to play tails.

Should first impulses prevail, you'll play heads and your opponent will play tails. There will be no match, and you will lose a penny to your opponent—hey, wasn't this game supposed to be stacked in your favor?

At a deeper level of analysis, you realize that your opponent pretty well *has* to pick tails. Not only does that veto your big win (his big loss), but he collects a penny every time you pick heads and he picks tails.

Two can play at that game. As long as you know your opponent is virtually certain to pick tails, you can take advantage of that fact. Choose tails, and you're almost sure to win a penny.

But maybe your opponent anticipates your double-cross. Then he might be tempted to play heads—or maybe not; he *is* risking a million that way. Still, if there's any chance at all that he might play heads, maybe you should reconsider playing heads. You can well afford to give up winning a penny for a long shot at the million. . . .

Game theory concludes that the correct mixed strategy is to play tails practically all the time. You should play heads with a probability of only about 2 in a 100 million (the exact ratio is 2:100,000,003.[3] Your opponent should do the same thing.

The million-dollar payoff, which seems to be such a windfall, is mostly an illusion since the other player can veto it. Regular matching pennies is a fair game with zero expected value. The millionaire version is in your favor, but only in the amount of approximately *one cent* per play. That, of course, is what you win for matching tails. The net effect of the million-dollar payoff is to raise your average gain by one cent! It would not change your expectation of gain appreciably if the bonus were raised to a trillion dollars or a googol dollars.

The other surprising thing is the recommendation that the second player occasionally play the risky strategy of heads. He doesn't play it *much,* but it is hard to rationalize playing it at all. Here's one way of looking at it. The game is basically one of both players playing tails

3. I'm not getting into the actual math here because it's not needed to understand social dilemmas. For "generalized matching pennies"—a two-person, zero-sum game with two strategies for each player—the correct mixed strategy is easy to calculate. Write the payoffs in a two by two grid as usual. Then calculate the differences of the two payoffs in each row and write them to the right of the table:

1,000,000	–	–0.01	=	1,000,000.01
–0.01	–	0.01	=	–0.02

Make both results positive (–0.02 becomes 0.02) and swap them:

1,000,000	–	–0.01	=	0.02
–0.01	–	0.01	=	1,000,000.01

This means the proper odds for heads:tails is 0.02:1,000,000.01, or (multiplying by 100 to get rid of the decimals) 2:100,000,001. The other player calculates his odds by figuring the differences in the columns and swapping. In this case the odds are the same for both players. It's more complicated for games with more than two strategies. If you're interested, see John Williams's *The Compleat Strategyst.*

(lower right cell). But were the second player to foreswear playing heads *completely,* that would rule out any possibility of your winning the $1 million. You would have no reason ever to play heads either.

The second player (who almost always plays tails) likes it when you play heads. That action almost always results in a win for him. He has to play heads occasionally to give you some incentive to keep on playing heads every now and then. What's more, these occasions when he plays heads usually turn out okay for him since you usually play tails.

Lightning rarely strikes twice. Provided both players play heads infrequently enough, the many, many cases of a single heads (and a penny's gain for the second player) balance the infrequent catastrophe of both players playing heads. Thus there is an optimal mixed strategy where heads are played very infrequently but not avoided entirely.

CURVE BALLS AND DEADLY GENES

Once you understand the idea of mixed strategies, you recognize them everywhere. Let's give a few examples.

Baseball pitchers are better at throwing some types of pitches than others. All other things being equal, the batter would expect the pitcher to throw his best pitch all the time. But should the batter know the type of pitch to expect, he would gain an advantage. Pitchers therefore throw a random mixture of fast balls, slow balls, curve balls, and knuckle balls to keep the batter uncertain. The rare exception only proves the rule. When Satchel Paige was asked how he could get away with always throwing fast balls, he answered, "They know what's comin', but they don't know where."

In principle, game theory could prescribe the optimal mixture of pitches. The mixture would vary according to the relative strengths of each player's pitches. You'd need some fairly exacting statistics—how many runs resulted from each type of pitch, ideally broken down by opposing batter. It would be interesting to see how closely pitchers' instinctive strategies approximate that of game theory. The math is no more involved than that in some of the baseball statistics being kept, and this seems a natural project for a future Bill James.

As early as 1928, Oskar Morgenstern recognized a dilemma in Arthur Conan Doyle's *The Adventures of Sherlock Holmes.* He and von Neumann cite it in their book:

Sherlock Holmes desires to proceed from London to Dover and hence to the Continent in order to escape from Professor Moriarty who pursues him. Having boarded the train he observes, as the train pulls out, the appearance of Professor Moriarty on the platform. Sherlock Holmes takes it for granted—and in this he is assumed to be fully justified—that his adversary, who has seen him, might secure a special train and overtake him. Sherlock Holmes is faced with the alternative of going to Dover or of leaving the train at Canterbury, the only intermediate station. His adversary—whose intelligence is assumed to be fully adequate to visualize these possibilities—has the same choice. Both opponents must choose the place of their detrainment in ignorance of the other's corresponding decision. If, as a result of these measures, they should find themselves, in fine, *on the same platform, Sherlock Holmes may with certainty expect to be killed by Moriarty. If Sherlock Holmes reaches Dover unharmed he can make good his escape.*

Von Neumann and Morgenstern go so far as to assign points to the various outcomes and compute a mixed strategy. They recommend that Moriarty go to Dover with a 60 percent probability, and to Canterbury with a 40 percent probability. Holmes should get off at Canterbury (60 percent probability) or Dover (40 percent probability). The game is unfair, and Moriarty has a better chance of prevailing.

In Doyle's story, Holmes gets out in Canterbury to see Moriarty's special train passing on its way to Dover. It is interesting that both Holmes and Moriarty followed the most likely course under von Neumann and Morgenstern's mixed strategy. They write, "It is, however, somewhat misleading that this procedure leads to Sherlock Holmes' complete victory, whereas, as we saw above, the odds (i.e. the value of a play) are definitely in favor of Moriarty. . . . Our result . . . yields that Sherlock Holmes is as good as 48% dead when his train pulls out from Victoria Station."

This kind of calculated deception resembles bluffing in poker. Poker can be quite complex, in part because it usually has more than two players. Von Neumann analyzed a simplified form of poker. In outline, his conclusions apply to the real game. He showed that you should always bid aggressively when you have a strong hand. With a weak hand, you should sometimes bluff (bid aggressively anyway).

Von Neumann distinguished two reasons for bluffing. A player who never bluffs misses many chances to call the other player's bluffs. Suppose that both you and your opponent have bad hands. You don't bluff; your opponent does. That means you fold and your opponent wins without a showdown. Had you also bluffed, your lousy hand would have been compared with his lousy hand, and you might have won. The bluffer can exploit the nonbluffer; ergo, von Neumann's rational player must bluff.

Bluffing is also a smoke screen. As in matching pennies, one wants to keep the other player guessing. Poker hands are dealt randomly to begin with, but players form opinions about their opponent's hands from their bids. Judicious bluffing prevents a player from being too predictable.

Game theory has important analogies in biology. A person who inherits the relatively rare sickle-cell anemia gene from one parent has greater immunity to malaria, but someone who inherits the gene from both parents develops sickle-cell anemia, a deadly disease. The puzzling survival of this and other lethal genes probably involves an equilibrium much like that in a bonus-payout version of matching pennies. In the latter game, a player picks the risky strategy of heads at rare intervals in order to get a benefit that accrues when just that player plays heads. The sickle-cell gene is likewise risky but confers a benefit when only one gene is present. Provided the gene is rare enough in the population, cases of the disease are rare compared to cases of enhanced immunity. This is believed to be the reason this seemingly unfavorable gene has persisted in areas where malaria is common.

You might wonder how this has anything to do with game theory. Genes can't choose mixed strategies or any kind of strategies. As it turns out, conscious choice is not essential to game theory. At the most abstract level, game theory is about tables with numbers in them—numbers that entities are efficiently acting to maximize or minimize. It makes no difference whether you picture the entities as poker players who *want* to win as much money as possible or as genes mindlessly reproducing as much as natural selection permits. We'll hear more about biological interpretations of game theory later.

THE MINIMAX THEOREM

The minimax theorem proves that *every* finite, two-person, zero-sum game has a rational solution in the form of a pure or mixed strategy.

Von Neumann's position as founder of game theory rests mainly with his proof of the minimax theorem by 1926. Von Neumann considered the theorem crucial. In 1953 he wrote, "As far as I can see, there could be no theory of games on these bases without that theorem. . . . Throughout the period in question I thought there was nothing worth publishing until the 'minimax theorem' was proved."

To put it in plain language, the minimax theorem says that there is always a rational solution to a precisely defined conflict between two people whose interests are completely opposite. It is a rational solution in that both parties can convince themselves that they cannot expect to do any better, given the nature of the conflict.

Game theory's prescriptions are conservative ones. They are the best a rational player can expect when playing against another rational player. They do not guarantee the best outcome possible. Usually a rational player can do *better* for himself when playing an irrational opponent. Sometimes these benefits accrue even to the rational player sticking with the prescribed strategy. In other situations it is necessary for the rational player to deviate from the strategy of game theory to take advantage of the other player's irrationality. An example is matching pennies. Say that you're the matching player and are mixing heads and tails equally and randomly. But you notice that your less rational opponent is unconsciously choosing "heads" more than half the time. Then you can come out ahead by choosing "heads" more often.

Sensible as this modification is, the modified strategy is no longer the optimal one and opens you to possible exploitation yourself (such as by a third player, or by the irrational opponent should he suddenly "wise up.")

N-PERSON GAMES

A journalist once asked von Neumann if game theory would help make a killing in the stock market. Honestly enough, von Neumann answered no. Such questions lingered. What *is* game theory good for? If not to play games, then what?

Von Neumann himself saw the minimax theorem as the first cornerstone of an exact science of economics. Toward this end, much of von Neumann and Morgenstern's book treats games with three or more persons. Most of the time, the number of "players" in an economic problem is large—huge even—and no simplifying assumptions are possible.

A game with an arbitrary number of players is called an "*n*-person game." A complete analysis of such games is much more complex than zero-sum two-person games. Conflicts of interest are less pat. What is good for player A may be bad for player B but good for player C. In such a situation, A and C might form an alliance. Such coalitions change a game radically.

In a three-person game, it is possible that two players acting in concert can guarantee a win. Two allies might thus cut a third player out of his share of winnings. Von Neumann and Morgenstern tried to decide when such coalitions were likely to form, and who was likely to form them. Would weak players gang up against a strong player? Or would weak players try to ally themselves with a strong player? One conclusion was that many potential coalitions can be stable. Then it is difficult or impossible to predict what will happen.

Von Neumann hoped to use the minimax theorem to tackle games of ever more players. The minimax theorem gives a rational solution to any two-person zero-sum game. A three-person game can be dissected into sub-games between the potential coalitions. If players A and B team up against player C, then the resulting game (coalition of A and B vs. C) is effectively a two-person game with a solution guaranteed by the minimax theorem. By figuring out the results of *all* the potential coalitions, the players A, B, and C would be able to decide which coalitions were most in their interests. This would then give a rational solution to a three-person game.

There's no need to stop there. A four-person game can be chopped up into two- and three-person games between its potential coalitions. Hash out all the possibilities there, and the solution will be evident. Four-person games lead to five-person games, six-person games, ad infinitum.

Unfortunately, the complexity of games, and of the necessary computations, increases exponentially with the number of players. If the economy of the world can be modeled as a 5-billion-player "game," that fact may be of little practical use. In the main, von Neumann and Morgenstern's work on economics never got off the ground. It remains for someone else to extend their foundations.

Good mathematician that he was, von Neumann did not try to limit his theory to its nominal subject matter. Geometry arose out of problems of surveying land. Today we find nothing unusual about using geometry in contexts that have nothing to do with real estate. A rectangle is a rectangle whether it's someone's farm or an abstract rectangle in a geometric proof. Von Neumann and Morgenstern point out that a zero-sum n-person game is in effect a function of n variables, or alternatively, an n-dimensional matrix. Much of the discussion in *Theory of Games and Economic Behavior* applies to such abstract functions or matrices irrespective of whether they are pictured as payoff tables for games, outcomes of economic or military decisions, or anything else. Game theory is inspired by, but not necessarily *about,* games.

Real conflicts postponed further development of game theory. Like many of his colleagues, von Neumann enlisted in the war effort. This left little time for pure research. Von Neumann would never again publish ground-breaking work in pure mathematics at the heady clip of the years between the world wars. Paul Halmos wrote (1973), "The year 1940 was just about the half-way point of von Neumann's scientific life, and his publications show a discontinuous break then. Till then he was a topflight pure mathematician who understood physics; after that he was an applied mathematician who remembered his pure work."

4

THE BOMB

During the war, von Neumann did consulting work for the Navy Bureau of Ordnance. Much of the time he was on alert to fly to England at a moment's notice. The strict baggage allowance was all the more restrictive due to a bulky antisplinter helmet he was supposed to pack. Von Neumann wanted to take along a volume of the *Cambridge Medieval History* to read instead. He took the helmet out of his packed bag and replaced it with the history. Klara dutifully took the book out and replaced it with the helmet. The switches went on for several months. Von Neumann won, leaving with the history when the call came.

Von Neumann spent the first half of 1943 in and around London. He wrote Klara (March 13, 1943): "Except for the blackout, which is also a thing one can get used to, life is *absolutely* normal here. . . . Alerts, air raids and the like are—certainly in the inner districts of London—just formalities."

The war put more strain on the von Neumanns' marriage. Letters between Johnny and Klara had to be written knowing they would be read by censors. Letters arrived late, out of sequence, or not at all. Klara argued with the Kupers over summer custody of Marina, angrily dropping the matter in the lap of her three-thousand-mile-distant husband. A considerable part of Johnny's letters is devoted to explaining unintentional or nonexistent slights Klara had found in previous letters. In one letter he goes to great lengths insisting he didn't mean anything when he compared Klara with a friend he described as having slightly dictatorial tendencies.

Klara complained of feelings of inferiority centering on her appearance and her accomplishments. These feelings had little basis in fact. In pictures, Klara is a handsome, stylishly dressed woman, certainly the more attractive of the couple. Though she felt she was in the shadow of her celebrated husband, it is also evident that Klara was a remarkably intelligent woman. During the war years, with Johnny away so much, she began to carve out a career for herself. She took a job with a population research study at Princeton. Though hired for her knowledge of languages, she excelled in the statistical side of the work.

VON NEUMANN AT LOS ALAMOS

From the very beginning, von Neumann was confident that the Allies would win the war. He actually sketched out a mathematical model of the conflict from which he deduced that the Allies would win, after a slow start, because of their industrial edge. One of von Neumann's wartime projects was a prime example of that industrial might, the atomic bomb. It was also to have a profound influence on the way game theory was perceived.

In late 1943, J. Robert Oppenheimer asked von Neumann to work on the Manhattan Project. Many scientists were reluctant to work on the bomb, doubting that such a farfetched project could be brought to fruition in time to affect the course of the war. Von Neumann, however, had for some years predicted that atomic energy would become a technological reality in his lifetime. The Manhattan Project beckoned with fresh mathematical challenges and an opportunity to serve his adopted country.

Von Neumann was one of the rare exceptions to the security rule that the scientists working on the bomb had to live in Los Alamos. Though other scientists were recruited to work on peripheral issues without moving to Los Alamos, they were not told that they were working on an atomic bomb. Von Neumann was judged so valuable to the project that he was permitted full knowledge of the bomb and freedom to come and go as his schedule allowed.

The concentration of great minds at Los Alamos underscored the remarkable fact that so many were from Hungary. The crowd of brilliant Hungarians—von Neumann, Edward Teller, Leo Szilard, Eugene Wigner, and Theodore von Kármán—were jokingly known as "the Martians." Stanislaw Ulam recalled (1958) that when von Neumann was asked about this "statistically unlikely" Hungarian phenomenon, von Neumann "would say that it was a coincidence of some cultural factors which he could not make precise: an external pressure on the whole society of this part of Central Europe, a subconscious feeling of extreme insecurity in individuals, and the necessity of producing the unusual or facing extinction."

Von Neumann did crucial calculations on the implosion design of the atomic bomb. The bomb's designers sought to create a critical mass of uranium or plutonium. In a critical mass, neutrons released split more atoms, releasing yet more neutrons to split more atoms. The chain reaction continues many times over, releasing immense energy in a short time. Creating a critical mass proved to be difficult. Theoretically, two hemispheres of uranium, each containing half a critical mass, need only be fitted together to create a critical mass. Normally, though, the burgeoning reaction would blow the two approaching hemispheres apart before they contacted each other.

It was clear that the uranium or plutonium had to be brought together very quickly. The atomic bomb dropped on Hiroshima ("Little Boy") used a "gun" trigger. It contained two pieces of uranium 235, a sphere with a hole in it, and a "bullet" that fitted into the hole. Neither piece was a critical mass itself, but the combination, with the bullet fitted into the sphere, was. The bomb mechanism consisted mainly of a gun that shot the bullet into the sphere with chemical explosives.

This arrangement was relatively inefficient. Another possibility was the implosion method, championed by Seth Neddermeyer. In this a hollow sphere of plutonium is surrounded with explosives confined by a strong outer shell. The explosives cause the plutonium to collapse to a much smaller, critical mass.

This idea also turned out to be much more complex in practice than conception, and it was here that von Neumann made important contributions. The imploding plutonium became a liquid. The implosion had to be almost perfectly symmetrical for the bomb to work. The joke was that it had to be like crushing a beer can without splattering any beer. Explosive "lenses," shaped charges containing both fast and slow explosives, were necessary to focus the imploding force on the plutonium. Von Neumann considered the problem in the fall of 1943, and his calculations led to a successful design for the lenses. One of von Neumann's important insights was that the implosion would compress the plutonium to greater than normal density. This would increase the speed of the chain reaction. Because of this phenomenon, the implosion bomb would produce a greater blast from a smaller amount of fissionable material. The "Fat Man" bomb that laid waste to Nagasaki was an implosion bomb.

GAME THEORY IN WARTIME

World War II was the first war in which game theory was put to practical use. During the war, mathematician John Tukey assigned Merrill Flood—who would later become co-discoverer of the prisoner's dilemma—to prepare a study on aerial bombing of Japan. The complex problem may have involved dilemmas similar to the Holmes and Moriarty problem in the previous chapter: If the United States *always* selected the most important target to bomb, the Japanese could anticipate that and have their antiaircraft forces defending it.

Flood recognized the game theory involved, as he had been one of von Neumann's students at Princeton. In fact, von Neumann had lent him an early version of his book's manuscript. Flood had had a not-entirely pleasant experience with game theory at Princeton. In 1939 he gave a lecture billed as "How to Win at Games of Chance." Flood used games such as craps to illustrate probability theory and then built up to game theory, using von Neumann's analysis of poker bluffing. At that time, poker was a popular pastime for Princeton students —too popular, some parents and trustees thought—and Flood was called into the dean's office to explain.

Using von Neumann's methods, Flood devised a bombing strategy that would minimize the chance of bombers being shot down. Because of wartime security, Flood didn't know exactly how this work would be used. Tukey wasn't allowed to tell him. Once Tukey hinted to Flood that it had something to do with a story in the paper about a mysterious flash being reported in the New Mexican desert. The study was, of course, for the Manhattan Project.

Von Neumann also provided strategic advice on the use of the atomic bomb, and helped General Leslie R. Groves select the locations in Japan to bomb. A sheet of notepaper dated May 10, 1945, now in the Library of Congress, has several lists of potential targets in von Neumann's handwriting. One runs: Kyoto, Hiroshima, Yokohama, Kokura. Kyoto was spared, according to histories of the war, because of its cultural importance.

Some associates felt that von Neumann ascribed game-theoretic rationality to the war's leaders. Valentine Bargman of Princeton recalled for Heims, "The von Neumanns were at our house one evening and I think the Paulis were there too. It was already clear that Hitler had lost the War. The question came up in our conversation, what will Hitler do? Johnny said, 'There is no question. The plane to South America stands ready.' "

According to Ulam, von Neumann felt that the defeat of Germany and Japan would lead to immediate war between the United States and the Soviet Union. Though war with the Soviets didn't follow the defeat of the Axis powers, he felt that war loomed as a possibility.

Unlike many of the Manhattan Project scientists, von Neumann found defense work stimulating. He seems not to have suffered the pangs of conscience that afflicted so many of the scientists who had worked on the bomb. Most of the Manhattan Project scientists returned to academia, many anxious to forget the war, the bomb, and Los Alamos. Von Neumann continued as a consultant to defense agencies after the war. He came to love the country around Los Alamos so much that some said that, had he lived, he might have bought a home there.

BERTRAND RUSSELL

The atomic bomb changed the face of war. Few realized that more than the British mathematician, philosopher, and mystic, Bertrand Russell (1872–1970). There are many parallels between John von Neumann and Bertrand Russell, despite more numerous and telling differences. Both men were among the most respected thinkers of their time, and both were, for a crucial phase of their careers, concerned with the axiomization of mathematics. In later life, von Neumann and Russell partially abandoned mathematics to spend a large part of their time on matters of war and peace. Russell was no game theorist, but he coined the name of the "chicken dilemma," one of game theory's most analyzed games.

Russell was born to an aristocratic family. At the age of eleven, he was severely disappointed to learn from his brother that the axioms of geometry could not be proven. He went to Cambridge's Trinity College, embarking on a career in mathematical logic.

Russell had a strong mystical streak—something not uncommon among mathematicians. In 1901, at the age of twenty-eight, he had a mystical experience that converted him to pacifism. Russell wrote in his autobiography, "Suddenly the ground seemed to give way beneath me, and I found myself in quite another region. . . . At the end of those five minutes, I had become a completely different person. For a time, a sort of mystic illumination possessed me. . . . Having been an imperialist, I became during those five minutes a pro-Boer and a pacifist."

During World War I, Russell remained true to his pacifism. He was fined for his promotion of antiwar views, and later sent to prison for six months. He also lost his position at Cambridge because of his pacifism.

In 1920 Russell visited the nascent Soviet Union and didn't like what he saw. He published a book (*The Practice and Theory of Bolshevism*) condemning the Bolsheviks' totalitarian rule. This was one of the first critiques of communism from a political liberal.

By middle age, Russell was a household word: philosopher as well as mathematician, and author of popular books on the theory of relativity, marriage, and education. He did not shy from publicity and was often quoted in the press. Russell spent the late 1930s and early 1940s in the United States. He was named professor at New York's City College, but this was revoked by a court order—partly because of his liberal views on sex. Russell saw nothing wrong with sex outside of marriage. Stranded without a job, Russell was rescued by Philadelphia millionaire Albert Barnes, who offered him a contract to lecture at the Barnes Foundation, a rather "experimental" art school and museum. Russell and Barnes clashed, and in 1944 Russell returned to England, securing a post at Cambridge.

After Japan surrendered, Russell believed, as von Neumann did, that war with the Soviet Union was inevitable. Just days after Hiroshima, Russell wrote an article in the *Glasgow Forward*.

> *Russia is sure to learn how to make it [the atomic bomb]. I think Stalin has inherited Hitler's ambition for world dictatorship. One must expect a war between U.S.A. and U.S.S.R. which will begin with the total destruction of London. I think the war will last 30 years, and leave a world without civilised people, from which everything will have to be built afresh—a process taking (say) 500 years . . . There is one thing and one only which could save the world, and that is a thing which I should not dream of advocating. It is, that America should make war on Russia during the next two years, and establish a world empire by means of the atomic bomb. This will not be done.*

On November 28, 1945, Russell gave a speech in the British House of Lords making many of the same points. "As I go about the street and see St. Paul's, the British Museum, the Houses of Parliament and the other monuments of our civilization, in my mind's eye I see a nightmare vision of those buildings as heaps of rubble with corpses all round them," he said.

Russell predicted bombs would get cheaper and proliferate. He even predicted that a fusion bomb would eventually be made. These prognostications may appear self-evident in retrospect, but at the time, many of those who had actually worked on the bomb were saying the opposite—that bombs would be horribly expensive to make for the foreseeable future, that the Soviets were decades away from their own bomb, and that a hydrogen bomb might never work.

WORLD GOVERNMENT

Russell was just one of a large number of people worldwide who started thinking about world government in the weeks after Hiroshima. In October 1945, Justice Owen Roberts gathered a distinguished group of scientists, literary figures, and statesmen at Dublin, New Hampshire, to christen an organization called World Government Now. An existing group, Americans United for World Organization, changed its name to the more explicit Americans United for World Government. The latter's avowed purpose was "the development of the United Nations Charter into a world agency adequate in delegated sovereignty to enforce the peace." A group at the University of Chicago took it upon themselves to pen a World Constitution.

Norman Cousins's book *Modern Man Is Obsolete* crystallized world government sentiments. Many other intellectuals, among them those in the "hard" sciences, were sympathetic. Astronomer Harlow Shapley and physicist Arthur H. Compton spoke in favor of world government. Albert Einstein, the most famous scientist of his time, called for a "supranational" body to govern atomic power. So did Edward Teller and Robert Oppenheimer, who usually appear as diametric opposites.

On October 16, 1945, the Army gave the Los Alamos staff a Certificate of Appreciation. In his acceptance speech, Oppenheimer said:

> *If atomic bombs are to be added to the arsenals of a warring world, or to the arsenals of nations preparing for war, then the*

time will come when mankind will curse the name of Los Alamos and Hiroshima.

The peoples of the world must unite, or they will perish. This war, that has ravaged so much of the earth, has written these words. The atomic bomb has spelled them out for all men to understand. Other men have spoken them, in other times, in other wars, or other weapons. They have not prevailed. There are some, misled by a false sense of human history, who hold that they will not prevail today. It is not for us to believe that. By our works we are committed, committed to a world united, before this common peril, in law and in humanity.

It wasn't just intellectuals who rallied to the world government banner. A number of political leaders talked guardedly along these lines—even President Truman. Shortly before the war ended, Truman said in Kansas City, "It will be just as easy for nations to get along in a republic of the world as it is for us to get along in the Republic of the United States."

For a time, world government was taken seriously enough to be debated in the U.S. Senate. On October 24, 1945, Senator Glen H. Taylor of Idaho introduced a resolution calling for a world republic.

Whereas the atomic bomb and other new and terrible instruments of warfare make it possible that most of mankind and civilization itself may be destroyed should the world become involved in another war; and

Whereas even before the soldiers of this war have returned to their homes another race between nations is already underway to train ever greater armies and to produce more scientifically diabolical weapons in the largest possible numbers . . . Now, therefore, be it

Resolved, *That the Senate of the United States hereby call upon the delegates of the United States of America to the United Nations Organization prayerfully and earnestly to redouble their efforts to secure world-wide agreement to:*

Limit and reduce immediately and eventually to abolish armaments, outlaw military training and conscription except for such police forces as the Security Council of the United Nations Organization may deem necessary to preserve the peace of the world; outlaw the manufacture or use of atomic bombs and all other atomic weapons for any purpose whatsoever; outlaw the manufacture or use of other weapons and instruments of war of every kind and nature, except for such weapons as the Security Council of the United Nations Organization may deem necessary to preserve the peace of the world . . . we therefore urge that every possible effort of our delegates to the United Nations Organization be directed toward the ultimate goal of establishing a world republic based upon democratic principles and universal suffrage regardless of race, color, or creed . . .

Taylor wielded little power in the Senate. He was the son of a minister and farmer in Idaho. He had achieved bemused press for having been an itinerant actor and a musician in a cowboy band before running for the Senate. Taylor's immodest proposal was the first resolution he had ever introduced.

Taylor worried that the constant threat of a surprise atomic attack would destroy morals. He told the Congress: "If a man feels that he may never live to see the dawning of another morning, probably he will decide to go out tonight and get drunk and celebrate and have a good time, and God knows what will happen to the world if such fatalistic attitudes become prevalent."

OPERATION CROSSROADS

Others were thinking about winning the next atomic war. One was Admiral Lewis Strauss. Like von Neumann, whom he befriended, Strauss achieved success early. Once a traveling shoe salesman, he became secretary to Herbert Hoover (then Food Administrator under Woodrow Wilson) at the age of twenty-one. Strauss became a partner in the New York investment banking firm of Kuhn, Loeb & Co. at thirty-three. He entered the Navy as a lieutenant commander and rose through the ranks to become rear admiral by the time of Hiroshima.

Peter Goodchild's *J. Robert Oppenheimer: Shatterer of Worlds* recounts this anecdote about Strauss. Attorney Harold Green had invited Strauss to a social occasion at his synagogue. Green said:

He walked into my house and kissed my wife who he'd never seen before and he said, "I understand you have children, may I see them?" And he heard them and without waiting for an answer he was in there playing with the children and changing the diapers of the baby. Later, I learned from my friend in division security that, before he accepted the invitation he had a very quick security check done on the Rabbi.

Green concluded ironically that Strauss "was a very human kind of person."

As the country returned to peacetime, Strauss worried about suggestions that the atomic bomb rendered navies obsolete. Strauss proposed a test of the effects of the bomb on warships. The idea was to assemble various types of empty ships in mid-ocean and drop a bomb nearby to see what would happen.

Many, including the Federation of American Scientists, opposed the plan. It had no scientific purpose, and the war was over. Though the test was ostensibly designed to help the U.S. Navy prepare for atomic attack, the United States was the only atomic power and presumably had no plans to bomb its own ships. The United States would be drawing conclusions about how foreign navies would fare under atomic attack. Indeed, the test used vessels captured from the German and Japanese navies.

Physicist L. A. Du Bridge of the University of Rochester—and formerly head of MIT's Radiation Laboratory during the war—wrote a letter objecting to the tests to the *New York Times*. After this letter was published, von Neumann and physicist Ralph Sawyer wrote a letter of rebuttal (May 7, 1946). They insisted the tests were "vital to increasing the defensive strength of the Navy."

Strauss's tests were dubbed "Operation Crossroads" and billed as the biggest scientific experiment in history. For much of the world, the tests were a gloomy demonstration that the bomb was not going to go away, even in peacetime.

Wild "end of the world" rumors preceded the tests. The *New York World Telegram* quoted Admiral W. H. P. Blandy, commander of Operation Crossroads, in rebuttal: "The bomb will not kill half the fish in the sea and poison the other half so they will kill the people who eat fish hereafter. It will not start a chain reaction in the water, converting it all to gas and letting ships on all the oceans drop to the bottom. It will not blow out the bottom of the sea, letting all the water run down the hole. It will not cause an earthquake or push up new mountain ranges. It will not cause a tidal wave. It will not destroy gravity." On the other side of the globe, a French designer appropriated the name of the test atoll for a swimsuit whose daring design seemed to symbolize the mad, modern age of the atom.

The tests were held July 1 and 25, 1946, on Bikini atoll in the Marshall Islands. They were the fourth and fifth times that atomic bombs had been exploded, and the first times such explosions were announced beforehand. About 40,000 people watched the Crossroads blasts, including a number of scientific, political, and press dignitaries. John von Neumann was among them. There were even two representatives of the Soviet Union.

The bomb, whose very size and shape was still a secret, was hidden behind a tent as it was loaded into the plane. *Time* magazine allowed that the bomb was big enough to have a foot-high pinup picture of Rita Hayworth pasted on its side.

The tests were covered live on radio for what has to rank as one of the strangest broadcasts ever. Microphones on the abandoned ships allowed listeners at home to eavesdrop on Armageddon. Mostly, they heard ducks. The doomed fowl, along with rats, pigs, and goats were left on some ships as part of the experiment. In an attempt to tailor the events to the demands of radio, a metronome was placed before an open mike on the *Pennsylvania.* Announcers asked the audience to concentrate on the tick: when the ticking stopped, it would mean that the bomb had exploded, and the metronome and microphone had ceased to exist.

Most of the witnesses were on ships some miles away. Observers ten miles away reported feeling heat like that from opening the door of an oven.

Von Neumann's consulting work took up an increasing share of his time. On November 18, 1946, von Neumann wrote Edward Teller, "I hope that the worst is over in your existence and that there is nothing ahead of you except the gay irresponsible life of Los Alamos . . . Seriously, I think that we all are completely crazy for leading the life we do, and you are probably worse than I am, although I don't quite know whether this statement is a sign of self-humility or of arrogance on my part."

THE COMPUTER

Of all von Neumann's postwar work, his development of the digital computer looms the largest today. It is probably through this that von Neumann's influence on the everyday lives of succeeding generations is greatest.

"In today's world, the unmodified word 'computer' normally refers to some version of the machine John von Neumann invented in the 1940s," Arno Penzias wrote in *Ideas and Information* (1989). If this overstates von Neumann's importance among the handful of computer pioneers, it is not by much. Blumberg and Owens' 1976 book quoted Edward Teller as saying that "probably the IBM company owes half its money to Johnny von Neumann." As adviser to IBM, von Neumann helped direct computer technology along the paths it took: stored-program (rather than hard-wired), digital (rather than analog) machines storing binary (rather than decimal) digits. Von Neumann and his colleagues did not pursue patents as aggressively as they might have, apparently in the interest of encouraging the technology.

The haphazard evolution of von Neumann's interest in computing machines runs something as follows. He had become interested in hydrodynamics and shock waves, partly in connection with the astrophysics of collapsing stars. The physics involved nonlinear effects that were poorly understood. As one of the experts on such effects, von Neumann was invaluable at Los Alamos. There he was impressed by the complexity of the calculations, and how slowly they progressed by hand and calculating machine. Also during the war, he studied the Army's ENIAC in Philadelphia and came up with ideas for a better computer.

After the war, von Neumann decided to build a new, better computer at Princeton. The idea got a lukewarm reception at the institute. Einstein remarked that a computer wouldn't help him devise a unified field theory. Von Neumann looked outside for funding. He and electrical engineer Vladimir Zworykin of RCA pitched the idea to the Navy's Lewis Strauss. They said the computer would predict the weather: hadn't the Normandy landing been imperiled by unpredictable Channel weather? The Navy, RCA, and other sources provided funds, and the machine was built in Princeton.

Princeton's computer was capable of a mere 2,000 multiplications a second. (Today, an IBM System 390 mainframe can handle about 41 million instructions a second.) Paul Halmos vouches for the following story, more amusing than amazing: when the time came to test the computer, someone proposed a suitable problem. In order to know whether the computer was working properly, they had to know the correct answer. This led to an impromptu John Henry contest in which von Neumann was pitted against the machine. Von Neumann got the answer first.

The Princeton computer didn't predict weather very well. Today's far speedier computers don't predict weather very well. But this and other early computers were applied to scores of applications more suited to their powers. The institute computer's first real application (mid-1950) was a set of calculations needed for the hydrogen bomb project.

Von Neumann's work on the early computers led to wide-ranging investigations of the possibilities of automata. Toward the end of his life, a desire for more efficient computers led him to a study of the human brain. In the preface to her husband's posthumously published *The Computer and the Brain,* Klara wrote that "until his last conscious hours, he remained interested in and intrigued by the still unexplored aspects and possibilities of the fast-growing use of automata." And though von Neumann himself might not have imagined it, computers would become important for basic research in game theory. Robert Axelrod's studies on iterated prisoner's dilemmas would have scarcely been feasible without computers.

Meanwhile, Klara became one of the first computer programmers. Coached by Johnny, she took on programming work for the Aberdeen Proving Ground and Los Alamos. Though she got this work through Johnny, a consultant to both facilities, the exacting nature of assembly language programming rules out any supposition of nepotism or dilettantism. Klara later called the computer work the most exciting thing she had ever done.

An odd rumor dogged von Neumann and the computer. The following, published in *World Digest* and credited to *Illustrazione del Popolo,* an Italian periodical, is typical of filler items that turned up in several nations in the late 1940s:

> *Professor Neumann, of Princeton University, U.S.A., has revealed that 40 per cent of the 800 scientists and technicians employed on the construction of "mechanical brains" have gone mad.*

The majority installed in American military mental homes, spend their time in prodigious arithmetical calculations.

A Canadian magazine editor clipped the item and addressed a letter to "Professor Neumann" at Princeton, where it was duly forwarded to John von Neumann. Von Neumann wrote back saying, of course, that it was a canard.

PREVENTIVE WAR

Meanwhile, Bertrand Russell took the step he himself had rejected, that of proposing a preventive war against the Soviets. Russell's first public call for preventive war seems to have been in the October 1945 issue of *Cavalcade,* a British popular magazine. Russell suggested that the Allies form a world federation and demand that the Soviet Union join: ". . . if the U.S.S.R. did not give way and join the confederation, after there had been time for mature consideration, the conditions for justifiable war, which I enumerated a moment ago, would all be fulfilled. A *casus belli* would not be difficult to find."

Privately, Russell confessed to a strong dislike for the Soviet Union. "I hate the Soviet Government too much for sanity," Russell wrote his friend Gamel Brenan in January 1946. In 1947, Russell wrote Einstein, "I think the only hope of peace (and that a slender one) lies in frightening Russia. . . . Generally, I think it useless to make any attempt whatsoever to conciliate Russia. The hope of achieving anything by this method seems to me 'wishful thinking.' "

On December 3, 1947, Russell gave a lunchtime talk to the Royal Empire Society in which he was more explicit:

I should like to see as soon as possible as close a union as possible of those countries who think it worth while to avoid atomic war. I think you could get so powerful an alliance that you could turn to Russia and say, "it is open to you to join this alliance if you will agree to the terms; if you will not join us we shall go to war with you." I am inclined to think that Russia would acquiesce; if not, provided this is done soon, the world might survive the resulting war and emerge with a single government such as the world needs.

Russell discussed these ideas with British military men. A letter written in May 1948 to Walter Marseille, who had proposed a plan for inspection of nuclear plants, stated:

> *During the past year, conversations with professional strategists have slightly modified my views. They say that in a few years we shall be in a better position, and that Russia will not yet have atomic bombs; that the economic recovery and military integration of Western Europe should be carried further before war begins; that at present neither air power nor atomic bombs could prevent Russia from over-running all Western Europe up to the Straits of Dover; and that the most dangerous period for us is the next two years. These views may or may not be correct, but at any rate they are those of the best experts.*
>
> *There are some things of which Europeans are more vividly conscious than Americans. If Russia overruns W. Europe, the destruction will be such as no subsequent re-conquest can undo. Practically the whole educated population will be sent to Labour camps in N.E. Siberia or on the shores of the White Sea, where most will die of hardships and the survivors will be turned into animals. (Cf. what happened to Polish intellectuals). Atomic bombs, if used, will at first have to be dropped on W. Europe, since Russia will be out of reach. The Russians, even without atomic bombs, will be able to destroy all big towns in England, as the Germans would have done if the war had lasted a few months longer. I have no doubt that America would win in the end, but unless W. Europe can be preserved from invasion, it would be lost to civilisation for centuries.*
>
> *Even at such a price, I think war would be worth while. Communism must be wiped out, and world government must be established. But if, by waiting, we could defend our present lines in Germany and Italy, it would be an immeasurable boon. I do not think the Russians will yield without war. I think all (including Stalin) are fatuous and ignorant. But I hope I am wrong about this.*

The press criticized Russell strongly for these views. On November 21, 1948, the *Reynolds News* complained of Russell, "The distilled essence of all the wisdom he has accumulated in a long life is this message of death and despair. Give up all faith in human reason, he tells us in effect. Resign yourselves to an endless orgy of killing, to the destruction of cities, to the poisoning of the fruitful earth by atomic radiation. Lord Russell, the famous philosopher, advances the oldest and most blood-drenched fallacy in History: 'the war to end wars.' "

Nevertheless, public attitudes about the bomb were changing. In the immediate aftermath of Hiroshima, there was a feeling that the bomb was too terrible to use. Ghastly photographs of radiation sickness victims and monstrous births demonstrated that the toll of suffering did not end with the mushroom cloud. But if the bomb was too terrible to use under any circumstances, then it had no deterrent value. In late 1948 U.S. Secretary of State George C. Marshall made the interesting comment that "until fairly recently I thought the Soviet leaders probably had felt that the American people would never permit the use of the bomb."

"Preventive war"—by that name—entered the American public consciousness in 1947, with the Soviet refusal to withdraw from East Germany. This solidified the American perception of the Soviet Union as no longer an ally but a nation unsympathetic to Western interests. U.S. Secretary of State James Byrnes suggested that the West use force to get the Soviet Union out of Germany. This led to speculative editorializing about an atomic attack on the Soviet motherland.

In its issue of October 31, 1947, *United States News* ran an article titled "Price of a 'Preventive War.' " "Talk of preventive war is publicly recognized—as an 'error'—by Henry L. Stimson, wartime Secretary of War," reported the magazine. "And, too, President Truman is hearing from at least one of his aides, who has been sounding popular sentiment, that people are saying it might be well to get war over with soon if there is to be a war."

Another leader who took preventive war seriously was the Commanding General of the Air Defense Command, George E. Stratemeyer. Until the UN could ensure peace, there could only be armistice in the "deadly game of war," he said in New York (December 11, 1947). Possibly game theory—certainly football—was on the general's mind when he warned, "Because of the fact that play may be resumed at any time, we must keep our winning team strong and intact—ready to take the field at a moment's notice. We are not blind as to whom the opposition may be, and shouldn't be blind as to the price we will have to pay if we should lose."

A small number of ordinary folks, frustrated and fearful, began thinking about using the bomb. "What Are We Waiting For?" is the

John von Neumann at afternoon tea, Institute for Advanced Study. (Photo by Alfred Eisenstaedt, *Life* magazine, © Time Warner Inc.)

Von Neumann lecturing to students. He was notorious for dashing out equations on the blackboard and erasing them before students had time to copy them. (Photo © Wolf-Semana/ *Time* magazine)

Merrill Flood.

Melvin Dresher.

Von Neumann and J. Robert Oppenheimer at unveiling of the Institute for Advanced Study's computer.
(Photo by Alan W. Richards, © Mrs. Alan W. Richards)

John and Klara von Neumann at their Princeton home. The dog is "Inverse."

Secretary of the Navy Francis Matthews. (Photo courtesy of the Harry S. Truman Presidential Library)

John von Neumann, confined to wheelchair, receives the Medal of Freedom from President Eisenhower. This was von Neumann's last public appearance. (Photo © UPI/Bettmann Newsphotos)

Robert Axelrod.

Bertrand Russell. (Photo © UPI/Bettmann Newsphotos)

musical question posed by Edwin Hopkins of New York. His 1948 pro-preventive war song asserts, "Oh, we've got the bombs to do the job/ Why let the despots thrive/ . . . Blast them to kingdom come/Let not the savage gang survive!"

5

THE RAND CORPORATION

The RAND Corporation is housed in a unspectacular low- and mid-rise complex a block from the beach at 1700 Main Street, Santa Monica. Blocky, in colors of pinkish terra-cotta and putty white, RAND's buildings resemble those that might occupy a California state college campus. Discrete signs identify the complex and warn that it is private property. There are no obtrusive gates or fences, and the landscaping of palms and large-leafed tropicals harks back to that favored in California in the 1950s. External louvered blinds project awkwardly in front of the windows to shield them from the sun. The windows of the higher floors overlook the Ferris wheel on the Santa Monica Pier, the Frank Gehry-designed Santa Monica Place mall, the civic auditorium, aging motels with names like the Flamingo West, and seafood restaurants adorned with giant clam shells and signs of anthropomorphic lobsters in chef's hats. Herman Kahn, one of RAND's best-known analysts, interrupted his thinking about the unthinkable to take a midday swim in the Pacific. When John von Neumann visited, he usually stayed in the nearby Georgian Hotel, still in business, now as an oceanfront home for senior citizens.

To many, the RAND Corporation epitomizes modern Machiavellianism. Both hawks and doves are apt to perceive it as a secret lair where amoral geniuses conspire darkly. RAND was well known enough to rate as a target for a Pete Seeger satirical folk song in the 1960s ("The RAND Corporation's the boon of the world/They think all day long for a fee./They sit and play games about going up in flames/For counters they use you and me . . ."[1]). *Business Week* reported, "The military professionals dub these civilian interlopers into the national security arena 'defense intellectuals,' 'RANDsters,' 'technocrats,' and worse. General Thomas D. White recently declared that in common with other military men, 'I am profoundly apprehensive of the pipe-smok-

1. "The Rand Hymn," words and music by Malvina Reynolds. © Copyright 1961 Schroder Music Co. (ASCAP). Renewed 1989 by Nancy Schimmel.

ing, trees-full-of-owls type of the so-called defense intellectuals who have been brought into this nation's capital.'" The general's apprehensions were shared by *Pravda,* which once called RAND the "American academy of death and destruction."

In his 1971 book, *Think Tanks,* Paul Dickson writes:

> *If the major problem in America today is misplaced priorities, then RAND is part of the problem. RAND, for example, has never worked on the very real problems encountered by the elderly today in America, but it has sternly considered the hypothetical problem of the elderly after nuclear war. . . . RAND suggested in 1966 that survivors of nuclear attack would be best off without the old and feeble, and U.S. policy should be to abandon them. Concludes the cold-blooded RAND report: "The easiest way to implement a morally repugnant but socially beneficial policy is by inaction. Under stress, the managers of post-attack society would most likely resolve their problems by failing to make any special provision for the elderly, the insane and the chronically ill."*

Public misgivings have not dampened RAND personnel's pride in their organization. In the preface to his ground-breaking 1960 book, *The Strategy of Conflict,* Thomas C. Schelling said, "As a collection of people, RAND is superb . . . But RAND is more than a collection of people; it is a social organism characterized by intellect, imagination, and good humor." Other insider views verge on the megalomanic. After the 1990 death of founder Franklin Collbohm, the *Los Angeles Times* quoted RAND spokesman Jess Cook as saying, "During its early years under Collbohm, it is not an exaggeration to say that RAND was one of the intellectual centers of the Western World."

HISTORY

The RAND Corporation derives from "operations research" conducted in World War II. As wars became both more terrible and more complex, it was realized that conventional military strategy was inadequate. Oskar Morgenstern writes in his 1959 book *The Question of National Defense:* "military matters have become so complex and so involved that the ordinary experience and training of the generals and admirals are no longer sufficient to master the problems. . . . The initiative to seek contact with science and scientists comes normally from the military men themselves. . . . More often than not, their attitude is: 'Here is a big problem. Can you help us?' And this is *not* restricted to the making of new bombs, better fuel, a new guidance system or what have you. It often comprises *tactical and strategic use* of the things on hand and the things only planned."

After the war ended, military leaders deplored the brain drain from the military back to universities and private industry. Military and civil service salary scales were unattractive to the most talented people. Few scientists wanted to work for the military, even at competitive pay. It was believed (incorrectly, as it turned out) that universities would be leery of taking on secret defense work. Several remedies were discussed. There was talk of a government "procurement agency" for brains. Another idea was contracting defense work out to private industry or hybrid military-industrial bodies.

In the summer of 1945, Douglas Aircraft sent Franklin R. Collbohm to Washington to lobby for the creation of a military research organization. Collbohm was a veteran engineer and test pilot who had helped design the classic DC-3 aircraft and had been co-pilot on its first flight. Army Air Force General Henry H. ("Hap") Arnold was particularly receptive to Collbohm's idea—not surprisingly, for one of General Arnold's sons was married to Donald Douglas's daughter, and Douglas's company had been a major supplier to the Army Air Force. Arnold decided to act on the idea on behalf of the Air Force.

On October 1, 1945, General Arnold met with Collbohm, Donald Douglas, Sr., and other Douglas Aircraft officials at Hamilton Field near San Francisco. Arnold personally committed $10 million of leftover defense funds to Douglas Aircraft for research. The $10 million was a great deal of money for the time; enough, it was hoped, to lure the most talented of scientists away from academia.

The legality of Arnold's move is debatable. Privileges of command are one thing, but this was $10 million of public money at a time when defense spending was winding down. While working for the military, Merrill Flood recalls coming across a red-bordered document marked "Eyes Only." His name wasn't on the list of those supposed to see it, but Flood leafed through it. It told of a research organization named "Project RAND" that was being funded with money intended for procurement. Flood showed the document to his superiors. It was expected that General Eisenhower would be furious.

The more Flood learned about the organization, the more it struck him as a good idea. RAND was engaged partly in defense applications of game theory, as in Flood's bombing study. Flood tried to convince his superiors of the project's merit. Eventually Eisenhower and the rest of the military command were sold on RAND, and the commitment to Douglas was honored.

Douglas executive Arthur Raymond coined the name "Project RAND," for "Research and Development." As the noncommittal name suggests, RAND's role was ill-defined. As originally formulated, the project would study intercontinental ballistic missiles (ICBM's). Bombers such as had been used at Hiroshima and Nagasaki would be too uncertain and vulnerable for future wars, it was felt.

Legally, RAND was a chimera, not quite a business concern and not quite a government agency. There was talk of Douglas creating a charitable foundation (!) to administer RAND. There was also some thought that RAND should be a project of the entire aerospace industry, not just Douglas. This led to the halfhearted creation of an advisory board of top executives of the other big aerospace firms.

Though supposedly accountable to no one but the Air Force, RAND was part of Douglas, housed on the second floor of Douglas's main offices in Santa Monica and headed by Collbohm, who still held his post at Douglas. Other firms feared that their private brainstorming with Air Force or RAND personnel would get back to Douglas Aircraft.

Progress on the secret work languished. The Air Force (which became a separate entity in 1947) grumbled that Project RAND wasn't attracting the best talent and even that Douglas was interested only in making a profit. In a letter dated September 5, 1946, Edward L. Bowles of the War Department wrote to General Arnold that "I have got the distinct impression that idealism has vanished from the picture and that we are working with Douglas on a strictly business basis, in which the Army Air Forces are underwriting any and all expenses of this project except what the Douglas Company may come by through taking over war assets."

Douglas Aircraft was none too pleased itself. In 1947 it lost an anticipated defense contract to Boeing. Many suspected the Air Force was bending over backward *not* to look as though it was favoring Douglas. The lost contract could have been more lucrative to Douglas than Project RAND would ever be. Douglas executives wondered if they hadn't succumbed to patriotic fervor without looking at the bottom line.

Finally, in February 1948, RAND's advisory council of aerospace executives (including representatives from Boeing, Northrop, and

North American) recommended that Douglas spin off RAND as a private, nonprofit firm. By then Douglas had no objection.

The RAND Corporation was chartered in March 1948. It became an anomaly among American institutions, a nonprofit body engaged in a lucrative "business" via a government contract. The Ford Foundation granted RAND a line of credit, and start-up capital was raised quickly. RAND's innocuous charter sounds almost as though it's describing the Smithsonian Institution: "to further and promote scientific, educational, and charitable purposes, all for the public welfare and security of the United States of America."

That November, the Air Force transferred the remainder of its contract with Douglas Aircraft (about half of the original $10 million) to RAND. For some years, the Air Force was its exclusive client.

RAND's Air Force contract granted it almost incredible freedom. The contract is somewhat less vague than the charter in calling for a "program of study and research on the broad subject of intercontinental warfare, other than surface, with the object of recommending to the Air Force preferred techniques and instrumentalities for this purpose." The "other than surface" clause was to prevent any appearance of usurping the Navy's territory. This contractual bailiwick was broader than its wording suggests. Truman's "atomic diplomacy" was predicated on the U.S. nuclear monopoly. Conventional forces were cut back. The Air Force, with its atomic capability, was the mainstay of the U.S. defense effort.

Within these loose guidelines, RAND's scientists were permitted to study any issue that interested them, with the Air Force footing the bill, regardless of whether the Air Force had any interest in the subject. Conversely, they could refuse specific studies requested by the Air Force (occasionally doing so for "nuts and bolts" research that failed to whet the intellectual palates of RAND's scientists).

In effect, RAND enjoyed a "grace period" in its early years in which it was not expected to produce much of anything. Hardly anyone knew what it was—not the public, not the press, not Congress. Many confused it with a typewriter company, Remington Rand. RAND was answerable primarily to General Curtis LeMay, Deputy Chief of Air Staff for Research and Development. According to Bruce Smith's *The RAND Corporation* (1966), LeMay "indicated that the new organization was to have a high degree of freedom to carry out its research objectives. Thus began a recurring pattern in RAND's history: RAND was fortunate enough to have 'protectors' at high levels within the Air

Force at crucial points to prevent it from falling victim to an internal power grab or from collapsing under severe budget cuts inspired by critics."

At least some of RAND's founders anticipated that it would manufacture weapons. Early on, RAND decided not to design actual hardware or even to do laboratory experimental work. A humbling incident got gossipy play in *Business Week* (February 8, 1947):

> *Project RAND, the Army Air Forces' supersecret Buck Rogers department, is due to be canceled. Chief reason: Now that it's ceased to be supersecret, it sounds a little embarrassing. Project RAND is a contract with the Douglas Aircraft Company to maintain a staff of assorted experts who spend their time looking into fantastic or sideline ideas occurring to AAF brass. The experts were getting top-priority on anything they wanted . . . RAND came under a cloud when a requisition on Oak Ridge for a supply of radioactive isotopes resulted in a War Department investigation to find out who wanted the stuff. Project RAND didn't get the isotopes.*

A long-standing organization joke claims that RAND stands for Research and No Development.

Internally, RAND structures itself more like a university than a military body or a corporation. Departments have academic titles like mathematics or environmental sciences. (Amazingly, it was once claimed that as many as 70 percent of new math Ph.D.'s applied for jobs at RAND.) RAND's ample military funding, with few strings attached, gave it a degree of freedom not unlike a well-endowed university. Several charitable bequests have helped fund research. RAND has held staff art exhibitions and concerts. The buildings are open around the clock to accommodate those who work eccentric hours. However, unlike a university, guards keep careful records of who visits and when they enter and leave. Though at present less than half of RAND's work is considered secret, the sensitive fraction mandates tight security.

RAND's busy publishing arm might be compared with a small university press. Hundreds of reports and books are issued each year. RAND published *A Million Random Digits with 100,000 Normal Deviates* (1955), a curious volume intended partly to facilitate the mixed strategies of game theory. One of the more popular RAND publications was John D. Williams's *The Compleat Strategyst* (1954). This was a "funny" primer on game theory for the interested layman, peppered with Williams's deadpan humor, RAND in-jokes, and cartoons. The book remains useful both as a readable, nonmathematical introduction and as a window on the RAND zeitgeist. A typical Williams observation (defending the use of random devices in mixed strategies): "A bomb isn't very intelligent either; for that matter the bombardier may on occasion give more thought to blondes than to target selection; of course, as we follow the chain back, it is comforting to suppose that pertinent intellectual activity occurs somewhere." The book was published in many countries, including the Soviet Union. In the Russian translation, a game of Russian roulette Williams analyzes was changed to "American roulette."

Like other organizations with secrets, RAND has been plagued by baseless rumors. On August 8, 1958, Senator Stuart Symington charged that the RAND Corporation was studying how the United States might surrender to an enemy power—something Symington felt they shouldn't be doing. Symington found in this a symbol of American defeatism. It turned out that Symington had either not read or grossly misunderstood the RAND study *Strategic Surrender.* The study was a survey of past cases in which the United States had demanded unconditional surrender of its enemies, asking whether this had been more favorable to U.S. interests than an earlier, negotiated surrender might have been. RAND's prompt explanation of the matter did not preclude two days' worth of Senate debate, leading to ratification of a law expressly forbidding the use of tax money to study defeat or surrender of any kind. This law still stands.

Apparently without any basis in fact was an April 1970 rumor, propagated by a Newhouse News Service story, that President Nixon had commissioned the RAND Corporation to study the feasibility of canceling the 1972 election. This was denied all around, and RAND even undertook a fruitless review of its recent work to see if there was anything that could have been misunderstood or distorted to spark the rumor.

Despite such conscientious policing of its image, RAND suffers from the perception of being frozen in the 1950s. It is now one of many such think tanks (spawned largely by RAND's success), and think tanks as an institution figure less prominently than they once did. In recent years, RAND has broadened its client base considerably. Its clients have included not only other branches of the Defense Department but also NASA, the National Institutes of Health, the Ford Foundation,

New York City, the state of California, and both the New York and American stock exchanges.

THINKING ABOUT THE UNTHINKABLE

In the public mind, RAND is best known for "thinking about the unthinkable," about the waging and consequences of nuclear war. Indeed, one of RAND's first projects was the selection of targets for a nuclear attack on the Soviet Union. In a memorable bit of RAND prose, theorist Herman Kahn asked (1960), "Would the survivors [of nuclear war] live as Americans are accustomed to living—with automobiles, television, ranch houses, freezers and so on? No one can say, but I believe there is every likelihood . . ."

The bomb predated intercontinental missiles by a decade. It is the ICBM as much as the bomb that creates the dilemmas of push-button war. Germany's V-2 missiles had a range of 200 miles, and that was considered astonishing at the time. The V-2's led to speculation about intercontinental missiles. In 1945 Vannevar Bush told the U.S. Senate, "There has been a great deal said about a 3,000 mile high-angle rocket. In my opinion such a thing is impossible for many years. The people who have been writing these things that annoy me, have been talking about a 3,000 mile, high angle rocket shot from one continent to another, carrying an atomic bomb and so directed as to be a precise weapon which would land exactly on a certain target, such as a city. I say, technically, I don't think anyone in the world knows how to do such a thing, and I feel confident that it will not be done for a very long period of time to come. . . . I think we can leave that out of our thinking." Bush, of course, was neither ill-informed nor lacking in imagination. Nor was von Neumann a few years later (December 1, 1948) when he wrote, in an unpublished letter to the editor of the *New Republic,* "In the case of the atomic bomb, I too feel that we are probably quite far from any form of 'push-button' warfare."

What changed things in a decade? One factor was the changing nature of the payload. The H-bomb was so much more powerful than the fission bomb that accuracy ceased to be a problem. A missile that "missed" and hit the suburbs would still wipe out the target city.

RAND and its extended family of consultants played crucial roles in developing the ICBM. (Intercontinental ballistic missiles briefly went by the acronym IBM. The computer company requested the change.)

Von Neumann became one of the most outspoken advocates for an ICBM program, despite his early skepticism. He came to speak of "nuclear weapons in their expected most vicious form of long-range missile delivery."

Bruno Augenstein, a RAND physicist, is credited with sketching out the intercontinental missile as it was actually built. RAND president Frank Collbohm took Augenstein's calculations to the Pentagon. In 1954 the United States decided to proceed with the Atlas ICBM program.

RAND studied disturbing or bizarre ideas that occurred to military leaders or to RAND's own thinkers. Some of RAND's researchers worried about the time delay in war, the necessary hours that elapse between a decision to launch an attack and the attack. It seemed to give the side that pushed the button first an overwhelming advantage. RAND devised the "fail-safe" protocol whereby bombers are kept in the air at all times. In times of crisis, they actually head toward targets in the enemy nation. Once they reach the fail-safe point they turn back unless they receive a go-ahead from the President.

One study asked, in effect: Suppose someone levels Cleveland; how would Washington find out, and how long would it take? RAND studied nuclear proliferation. Who would get the bomb and how soon; how cheap, portable, and easy-to-use could an atomic bomb be? One semi-serious RAND idea was Californium bullets. Each bullet is a barely subcritical mass of the highly fissile isotope. Pop one into a heavily shielded, long-range rifle, and when the bullet hits its target, it explodes with the force of tons of TNT.

In still another study (initiated by RAND and not the Air Force) RAND contemplated the likelihood of accident, sabotage, or psychotic Air Force personnel launching an unauthorized nuclear strike. RAND concluded that all three scenarios were plausible enough to take seriously; in fact, it would be quite possible for an irrational individual to start a nuclear war provided he was in a position of responsibility. The Air Force adopted RAND's suggestions for better psychological screening of those working with the bomb and for the "permissive action link," the safer "button" that requires the cooperation of several individuals to arm and detonate nuclear warheads.

In 1951 the Air Force asked RAND to propose sites for new European bases. The project was assigned to mathematician and economist Albert Wohlstetter, who nearly turned it down as too mundane. "It looked to me like a very dull problem in logistics; hairy but uninteresting," he was quoted as saying in Harpers (1960). Wohlstetter agreed to take on the project on the condition that he could broaden it to include abstract questions of deterrence.

RAND's 1954 report was one of the most influential in forming public policy. The report started by explaining that the Air Force had been wrong to ask its question in the first place. According to the study, overseas bases were not cost-effective and were sitting ducks for a Soviet surprise attack. Instead, the report suggested building more domestic bases and a strategy of "second-strike" capability. This was defined as the ability of the United States to launch a counterattack on the Soviet Union even in the aftermath of a Soviet first strike that had killed most Americans. Second-strike capability has been a cornerstone of Pentagon thought ever since. RAND supported the Polaris submarine program as conducive to second-strike capabilities. Since nuclear submarines do not stay in one place, it is nearly impossible for an enemy to destroy them all and prevent a counterattack.

RAND has done work on disarmament, too. One of the best-known studies was discouraging. RAND physicist Albert Latter did a study concluding that the Soviets could disguise underground nuclear tests by making them produce the same type of shock waves as an earthquake.

SURFING, SEMANTICS, FINNISH PHONOLOGY

For years, John Williams (who would be one of the subjects in the first prisoner's dilemma experiment) supervised RAND's mathematics department. Williams, born to a wealthy family, was trained in mathematics and was an enthusiast in many fields—among them meteors. Under Williams, a study of one issue led to a network of related issues, and RAND's roster of experts became increasingly diverse.

By 1960, RAND boasted a staff of 500 full-time researchers and about 300 outside consultants. RAND studied math education, neurosis, and class systems in Arab politics. Paul Dickson enumerated some of the "work of little or no consequence to the world at large" done at RAND: "the price of bricks in the Soviet Union, surfing, semantics, Finnish phonology, the social groupings of monkeys, and an analysis of the popular toy-store puzzle 'Instant Insanity.' "

It is easy enough to cry "golden fleece" and make almost any type of research sound silly. RAND supporters point out that many of these studies paid off with unexpected benefits. One was the space program. In the years before NASA, RAND was the principal U.S. organization engaged in space research, thanks to the permissive terms of the Air Force contract. Many of the problems RAND tackled for the ICBM project, such as designing a nose cone capable of reentry, were no less applicable to peacetime space flight. In 1946, RAND published a "Preliminary Design of an Experimental World-Circling Spaceship." A decade before Sputnik, this report claimed:

> *The achievement of the satellite craft by the United States would inflame the imagination of mankind, and would probably produce repercussions in the world comparable to the explosion of the atomic bomb . . . Since mastery of the elements is a reliable index of material progress, the nation which first makes significant achievements in space travel will be acknowledged as the world leader in both military and scientific techniques. To visualize the impact on the world, one can imagine the consternation and admiration that would be felt here if the U.S. were to discover suddenly that some other nation had already put up a successful satellite.*

The social sciences became important at RAND. Williams met with General LeMay in late 1946 to get support for hiring social scientists. Williams told Bruce Smith, "They sent me to Washington to kill the idea so I'd stop pestering people." LeMay was skeptical at first. Williams convinced him that it was important to understand the Soviets and that spending a small amount of money on social sciences could save greater amounts elsewhere. At the end of the meeting, Williams cautiously asked LeMay if he understood him correctly, that he was okaying the hiring of a few social scientists. LeMay said, "No, no, that's not it. Let's do it up right. If we're going to do this, do it on a meaningful scale."

One consultant was noted philosopher Abraham Kaplan. Melvin Dresher recalls that Kaplan was on a cross-country flight when his seatmate asked him what company he worked for. RAND was still part of Douglas, so Kaplan said he worked for Douglas Aircraft. "How's business?" the other man asked. "I don't know," Kaplan confessed. "What do you do for Douglas?" the man wanted to know. "I'm a philosopher," Kaplan said.

Judging it difficult to gauge Soviet intentions from official sources of information, RAND made an almost fantastic effort to get inside the Soviet mind. RAND instituted a hermeneutic study of the writings of Lenin and Stalin (*The Operational Code of the Politburo*) in the hope that this would assist American diplomats in dealing with their Soviet colleagues. They commissioned no less a figure than Margaret Mead to do a study on Soviet attitudes toward authority. One of the most amazing of the Soviet studies was the creation of a doppelgänger Soviet Ministry of Economics in Santa Monica. Using documents captured from the Nazis, RAND created a detailed model of the economy of the U.S.S.R.

Many of the social scientists hired or consulted were economists who had fallen under the spell of game theory. It was at RAND rather than in the groves of academia that game theory was nurtured in the years after von Neumann and Morgenstern's book. In late 1940s and early 1950s, few of the biggest names of game theory and allied fields *didn't* work for RAND, either full-time or as consultants. Besides von Neumann, RAND employed Kenneth Arrow, George Dantzig, Melvin Dresher, Merrill Flood, R. Duncan Luce, John Nash, Anatol Rapoport, Lloyd Shapley, and Martin Shubik—nearly all of whom were there at the same time. It is difficult to think of any other scientific field in which talent was concentrated so exclusively at one institution. Most of the above workers had left by 1960, but the RAND diaspora continued to dominate game theory throughout the academic world.

VON NEUMANN AT RAND

John von Neumann's formal association with the RAND Corporation began in 1948. On December 16, 1947, Williams, then of Project RAND, wrote offering von Neumann a retainer fee of $200 a month for his services. "In practice I would hope," Williams wrote, "that members of the Project with problems in your line (i.e., the wide world) could discuss them with you, by mail and in person. We would send you all working papers and reports of RAND which we think would interest you, expecting you to react (with frown, hint, or suggestion) when you had a reaction. In this phase, the only part of your thinking time we'd like to bid for systematically is that which you spend shaving: we'd like you to pass on to us any ideas that come to you while so engaged."

In a 1948 letter, Williams promised von Neumann: "We intend to make major efforts on applications of game theory . . . If you could spare some time for us during the summer, particularly during July and August, it would vastly stimulate the work here and, I believe, interest you too. . . . If you were really to pour your torrent of energy into these subjects for a while, there would probably be a handsome pay off."

Von Neumann enjoyed the RAND milieu. John Williams lived in a house in Pacific Palisades built by a millionaire of a previous generation. The original house had been too big to sell at the owner's death—no one was willing to pay a price commensurate with its square footage. A developer got the bright idea of cutting the house into five rectangular slices and demolishing the second and fourth slices. This produced three houses in place of one. Williams bought the middle house—actress Deborah Kerr lived in one of the others—and he threw the sort of high-proof, high-I.Q. parties that von Neumann was also known for.

At one of these parties, Merrill Flood wanted to show von Neumann the "three-sided coin" that had been a wry obsession at RAND. Someone had proposed that a thick cylindrical "coin" of the proper dimensions would fall heads one third of the time, tails one third of the time, and on its side one third of time. RAND scientists promptly set about calculating the necessary dimensions. Williams was so taken with the idea that he had a machine shop mill a few of the "coins" as jokes. While talking with von Neumann, Flood found a pretext for needing a three-way random choice. Let's flip a coin, Flood said. No good, there are three possibilities, von Neumann reminded him. Flood coolly produced the coin. Von Neumann looked at it, thought a moment, and announced the coin's proportions. He was correct.[2]

Another von Neumann "infallibility" story: Supposedly, RAND was working on a problem so complex that no existing computer could handle it, and they asked von Neumann's help in designing a new, more powerful computer. Von Neumann asked that they first tell him the problem. The scientists' explanation took about two hours, accompanied with furious scribbling on a blackboard. Von Neumann just sat there with his head buried in his hands. At the end of the explanation, von Neumann scribbled something on a pad of paper in front of him.

2. But Flood doubts that even von Neumann could have calculated that fast. He suspects that someone had told von Neumann of the coin earlier.

"Gentlemen," he said, "you do not need a new computer. I have just solved the problem."

Von Neumann's position at RAND meant that he was employed on both coasts—and RAND was not the only outside job he had. His constant travel exacerbated his marital troubles. Klara complained that she was unappreciated and that Johnny was interested only in work—doubtless with reason. She would refuse to accept her husband's long-distance phone calls. In one letter (May 2, 1949), Johnny reiterates that he loves and misses Klara, but tells her not to expect constant proof of his feelings. No one this side of Elysium, he says, can demonstrate love all the time without being boring.

JOHN NASH

After von Neumann, the next major figure in game theory was another RAND consultant, John F. Nash. Nash was born in Bluefield, West Virginia, in 1928. He studied mathematics at Princeton University. There Nash became interested in games. In 1948 he devised a game that was played with markers on a diamond-shaped board of hexagonal cells or on hexagonal bathroom tiles. This game quickly became a fad at Princeton and at the Institute for Advanced Study. It was called either "Nash" or "John," the latter also referring to the fact that it was played on bathroom floors. (Essentially the same game had been popular at Copenhagen's Bohr Institute since 1942, but Nash seems to have been unaware of it.)

As with chess, there is a correct way to play Nash's game, but it is impractical to discover it. Nash proved, however, that the correct strategy must lead to a win for the first player. A version of the game was marketed as "Hex" by Parker Brothers in 1952.

In due course Nash followed von Neumann's example by juggling a bicoastal career as RAND consultant and professor at the Massachusetts Institute of Technology. In the late 1940s and early 1950s, Nash extended game theory in a direction von Neumann and Morgenstern had not taken it. Nash studied "noncooperative" games where coalitions are forbidden.

Von Neumann and Morgenstern's treatment of games of more than two persons focuses on coalitions, groups of players who chose to act in concert. They suppose that rational players would hash out the results of joining every possible coalition and choose the one most advantageous. This approach makes sense given von Neumann and Morgenstern's grand aim, which was to treat economic conflicts as *n*-person games. Businesses team up to fix prices or drive a competitor out of business; workers join unions and bargain collectively. In each case it is reasonable to expect that parties will form coalitions whenever it is to their advantage. In effect, this is the definition of a free-market, *laissez-faire* economy.

The only kind of noncooperative games von Neumann treated were two-person, zero-sum games—which are necessarily noncooperative. When one player's gain is another's loss, there is no point in forming a coalition. That case, however, was already covered by von Neumann's minimax theorem. Nash's work was primarily concerned with non-zero-sum games and games of three or more players.

With the minimax theorem, von Neumann struck a great blow for rationality. He demonstrated that any two rational beings who find their interests *completely* opposed can settle on a rational course of action in confidence that the other will do the same. This rational solution of a zero-sum game is an equilibrium enforced by self-interest and mistrust—and the mistrust is reasonable in view of the antithetical aims of the players.

Nash extended this by showing that equilibrium solutions also exist for *non-zero-sum* two-person games. It might seem that when two persons' interests are *not* completely opposed—where by their actions they can increase the common good—it would be even easier to come to a rational solution. In fact it is often harder, and such solutions may be less satisfying.

THE MONDAY-MORNING QUARTERBACK

The heart of Nash's analysis is delightfully simple. We have all listened to Monday-morning quarterbacks tell how they would have played a game had they been in charge. "I would have concentrated on passing, and the Redskins would have won the game!"

There is one rule implicit in such free fantasizing. You can't change the *other* team's strategy. If you're talking about what the Redskins should have done, you can't also presume to change the way their opponents played. That would make it too easy. If you could choose the opposing team's strategy, you could sabotage their play. That's not fair.

Nash's approach to noncooperative games emphasizes "equilibrium points." These are outcomes where the players have no regrets. Hold a postmortem analysis after the game. Go to each player in turn and ask him if he would have done things any differently *given* how the other player(s) played. If everybody is happy with the way they played, then that outcome is an equilibrium point.

Here's an example of a non-zero-sum game and its equilibrium point solution:

	Strategy 1	Strategy 2
Strategy 1	1, 100	0, 1
Strategy 2	2, 0	**5, 2**

We can use the same type of table we have been using for zero-sum games, with one difference. It's necessary to put two numbers in each cell of the game table. The first gives the payoff to the "row player" (the one who chooses the row of the outcome). The second number in each cell is the payoff of the "column player." No longer can we assume that one player's gain is another's loss. Some cells have a higher combined payoff than others.

The Nash equilibrium solution is for both players to choose their strategy 2 (lower right cell, boldface). Obviously the row player is satisfied with this, for he wins 5 points, the most he could win under any circumstances. But this outcome can be justified to the column player as well. Playing Monday-morning quarterback, *given* that the row player chose strategy 2, the column player cannot regret having chosen his strategy 2. Had he chosen strategy 1, he would have won nothing at all. At least this way he wins 2 points.

Yes, but what's wrong with the upper left cell, the column player might ask (there he wins 100 points!). The answer is that it is an unrealistic outcome because the row player can't justify it. Suppose both players *did* choose their strategy 1. In the postmortem analysis, the row player would conclude he would have been better off choosing strategy 2 (2 points rather than 1). Nash argued reasonably that any outcome where a player would change his strategy, given the chance, is unstable and, presumably, not an example of rational play. Of the four outcomes, only the lower right one leaves both players with no regrets.

This sounds like a reasonable description of a "rational solution." Nash proved that *every* two-person finite game has at least one equilibrium point. This result is an important extension of von Neumann's minimax theorem. The minimax solutions of zero-sum games qualify as equilibrium points, but Nash's proof says that non-zero-sum games have equilibrium points, too. That is a new result.

But there are a few catches. These equilibrium points can have "strange and undesirable properties," as Philip D. Straffin, Jr., put it (1980). The above example was chosen to show a game where the equilibrium point solution clearly makes sense. Other times, equilibrium point solutions appear less inevitable than the solutions of zero-sum games. In fact, sometimes Nash equilibriums appear to be distinctly irrational. We will explore the consequences of this in the following chapters.

6

PRISONER'S DILEMMA

People are irrational. RAND's Merrill Flood was not the first to realize that, but he was one of the first to analyze that irrationality with game theory. Beginning in 1949, Flood sought out interesting games, dilemmas, or bargaining situations in everyday life. He asked the people involved how they had decided what to do. Were they (unconsciously!) using the von Neumann-Morgenstern theory, the Nash equilibrium theory, or something else entirely? Flood also accumulated data on how departing RAND colleagues sold or gave away their belongings (many stayed just for the academic summer break). One consultant donated the leftovers of his summer in Santa Monica —"a fifth of a fifth of Scotch whiskey, half a box of prunes, seven eggs, a dilapidated suitcase, some kitchen utensils, and odds and ends"—for an experiment on the mathematical "fair division" theory of economist Hugo Steinhaus. Flood reported several of these investigations in "Some Experimental Games," a RAND research memorandum dated June 20, 1952.

THE BUICK SALE

In June 1949, Flood wanted to buy a used Buick from a RAND employee who was moving back East. Buyer and seller were friends. They weren't looking to cheat each other, just to agree on a fair price for the car. How should they set a price?

As it happened, Flood and the seller knew a used-car dealer. They took the car to the dealer and asked him his selling and buying price for the car in "as is" condition. The difference, the dealer's profit, was a gain the buyer and seller could split between themselves.

Let's say the dealer's buying price was $500, just to have a concrete figure. The seller could, if he wanted, sell the car to the dealer for that price. Likewise, the buyer could buy a car just as good as the Buick from the dealer for the dealer's selling price—say, $800. In a transaction handled by the dealer, the dealer's cut would be $300. By not going through the dealer, the buyer and seller have an extra $300 to split between themselves.

How should buyer and seller split the $300 profit? They could divide it right down the middle. The sale price would be the dealer's buying price of $500 plus half of the $300 profit, or $650. The seller would get an extra $150 for his car, and the buyer would get an $800 car for only $650.

That sounds fair. In point of fact, it's what the two RAND employees did. The only thing is, it's not a unique solution. Both buyer and seller are in a position to veto *any* price. Should either party want to make an issue of it, they could demand a different split.

The buyer could be obstinate and insist that he won't pay any more than $600 . . . or $550, or $525, or even $501. The owner could tell him to go take a hike, but still, if the owner goes to the dealer, he'll only get $500. He's hurting himself by not accepting the buyer's offer, no matter how low (in excess of the dealer's price) it is.

It works the other way, too. The seller can be just as obstinate and name a price near the dealer's selling price. The peculiar thing is that the party who is more *unreasonable* is apt to get the better of the deal. This isn't news to used-car dealers, but it's a little disturbing.

The underlying thread connecting many of Flood's observations and experiments was "splitting the gain." When people can cooperate to secure an extra gain, how do they split it among themselves? Flood cooked up what he thought was a pretty good experiment. He offered two RAND secretaries the following deal: he would either give the first secretary a cash prize (say, $100) *or* give both secretaries a larger prize (say, $150) *provided* they could agree how to split the larger amount between themselves and would tell Flood their reasoning.

This experiment differs from the Buick sale because the first secretary alone is privileged and can secure the $100 without any assistance from the other. The other secretary is guaranteed nothing unless the first secretary cooperates. The problem, Flood assumed, was divvying up the extra $50. He supposed that they would split the difference, as in the car sale. The privileged secretary would get $125 and the other, $25. The secretaries didn't see it that way. They agreed to split the total $150 down the middle! Flood concluded that the social relationship of the parties made a big difference in how they acted.

Even kinship was no guarantee of cooperation, however. Flood needed one of his three teenaged children to baby-sit and held a "reverse auction" for the job. The child agreeing to baby-sit for the *lowest* pay would get the job. The opening bid was set at $4.00. Flood encouraged the children to come to an agreement among themselves to avoid a bidding war (this is what the von Neumann-Morgenstern theory of *n*-person games assumes). Despite the fact that the children were given several days to confer, they came to no agreement and bid against each other. The job went for a low bid of 90 cents.

Flood noted, "This is probably an extreme example, although not really so extreme when the magnitude of the children's error is compared with that made by mature nations at war because of inability to split-the-difference. I have noticed very similar 'irrational' behavior in many other real life situations since August 1949 and find it to be commonplace rather than rare."

HONOR AMONG THIEVES

Of the practical dilemmas in Flood's paper, by far the most important was the third, dubbed "A Non-cooperative Pair." The first scientific discussion of a prisoner's dilemma, this part of the paper describes an experiment done in January 1950 in collaboration with RAND colleague Melvin Dresher.

The original experiment is probably not the best way to introduce the prisoner's dilemma. Instead, let's jump to a modern version of the prisoner's dilemma, in the form of a story.

Suppose you have stolen the Hope Diamond and are trying to sell it. You learn of a potential buyer, an underworld figure called Mr. Big—the most ruthless man on earth. Though extremely intelligent, Mr. Big is notoriously greedy and equally notorious for double-crossing. You have agreed to exchange the diamond for an attaché case full of $100 bills. Mr. Big suggests that you meet out in a deserted wheat field somewhere to make the exchange. That way there are no witnesses.

You happen to know that Mr. Big has negotiated with many other sellers of contraband in the past. Each time he has suggested a remote locale for the exchange. Every time, Mr. Big showed up and opened the attaché case to show his good will. Then Mr. Big pulled out a tommy gun, shot the other person dead, and left with both the money *and* the goods.

You say you don't think the wheat-field plan is such a good idea.

You suggest the two-wheat-field plan. Mr. Big hides his attaché case of money in a wheat field in North Dakota, while you hide the diamond in a wheat field in South Dakota. Then both parties go to the nearest public phone and exchange directions on how to find the hidden goods.

This plan has built-in safeguards (tactfully, you don't mention that). You need have nothing of value on you when you go to recover Mr. Big's attaché case. Mr. Big (who is no homicidal maniac, just a shrewd businessman) will have no reason to wait in the North Dakota field to ambush you. Mr. Big agrees to the two-wheat-field plan.

You find a wheat field in South Dakota. As you are about to hide the attaché case with the diamond, an idea pops into your head. *Why not just keep the diamond?* Mr. Big will have no way of knowing that you betrayed him until he gets to South Dakota (you would wait for his phone call and give him directions as if nothing were wrong). By that time, you would be in North Dakota to pick up the money. Then you would hop on a plane to Rio. You would never see Mr. Big again.

A worse thought pops into your head. Mr. Big must be thinking the exact same thing! He's just as smart as you are, and probably ten times greedier. He has equal incentive to betray you, and you wouldn't be able to retaliate any more than he will.

The dilemma looks like this:

	Mr. Big sticks to agreement	**Mr. Big cheats**
You stick to agreement	Deal goes through: you get money, Mr. Big gets diamond	You get nothing, Mr. Big walks away with diamond and money
You cheat	You walk away with money and diamond, Mr. Big gets nothing	A lot of trouble for nothing: you keep diamond, Mr. Big keeps money

The problem is that you have to make a decision in ignorance of Mr. Big's decision and then live with it. You would most prefer to get the money without surrendering the diamond. Mr. Big would most prefer to get the diamond for nothing. Make no mistake about it, though, you both would be sincerely happy to have the deal go through to the letter of the agreement. Mr. Big really wants that diamond for his trophy case—not just any diamond, but the one-and-only *Hope Diamond.* He knows that you're his only chance to get it. Likewise, you really want the money. Mr. Big has offered a fabulous price, far more than anyone else would.

The best outcome all around is the upper left cell—the result when both comply with the bargain. But the best outcome for any individual is to be the lone cheater. The *worst* outcome is to be a sucker who sticks to the agreement while the other person cheats.

Here's one way of looking at it. Your actions in South Dakota cannot possibly influence Mr. Big's actions in North Dakota. No matter what Mr. Big does, you are better off keeping the diamond for yourself. If Mr. Big leaves the money, you end up with the money and the diamond. If Mr. Big leaves nothing, at least you still have your diamond to sell to someone else. So you should cheat and leave nothing.

Here's another way of looking at it. You're both in the same boat. Take the reasoning of the previous paragraph a step further. Mr. Big is perfectly capable of coming to the same conclusion, that it is "rational" to cheat. Then both of you will cheat, and both will go to a lot of trouble for nothing. Logic (?) blocks a deal beneficial to both parties. There's nothing logical about that! Therefore you should stick to the agreement. You should be sensible enough to realize that cheating undermines the common good.

This is a prisoner's dilemma, and now is a good time to ask yourself, what would you do?

This formulation of the dilemma was popularized by cognitive scientist Douglas Hofstadter. Here the dilemma is particularly easy to appreciate. Even among the law-abiding, most transactions are potential prisoner's dilemmas. You agree to buy aluminum siding: how do you know the salesman won't skip town with your down payment? How does he know you won't stop payment on the check? In my grade school, the accepted way to swap toys was for each child to set his toy down on the ground in plain sight some distance from the other, and then to run to the other toy. (If they just handed the toys over, one child could grab both.) With this arrangement, each child could see that the other had surrendered his toy, avoiding the cheating dilemma above! The slightly more adult equivalent of this is the escrow in real estate transactions. And speaking of crime, newspaper accounts of soured drug deals often report that someone tried to cheat more or less as above (not always with impunity).

THE FLOOD-DRESHER EXPERIMENT

Flood and Dresher were concerned that Nash's equilibrium point solutions could be unsatisfying. Remember, an equilibrium point is an outcome in which no one, playing Monday-morning quarterback, regrets choosing the strategy he did *given the other player's choice*. But there can be cases where the equilibrium point isn't such a good outcome.

Flood and Dresher devised a simple game where that was the case. The researchers wondered if real people playing the game—especially, people who had never heard of Nash or equilibrium points—would be drawn mysteriously to the equilibrium strategy. Flood and Dresher doubted it.

The researchers ran an experiment that very afternoon. They recruited two friends as guinea pigs, Armen Alchian of UCLA ("AA" below), and RAND's John D. Williams ("JW"). The game was presented purely as a payoff table. The payoffs were:

	JW's Strategy 1 [Defect]	**JW's Strategy 2 [Cooperate]**
AA's Strategy 1 [Cooperate]	–1¢, 2¢	1/2¢, 1¢
AA's Strategy 2 [Defect]	0, 1/2¢	1¢, –1¢

Don't worry about the values of these payoffs. They are purposely a little confusing in order to conceal the equilibrium point.

Each player was required to choose his strategy in ignorance of the other player's choice. If Alchian chose his strategy 1 (upper row) and Williams chose his strategy 1 (left column), then Alchian would be penalized a cent and Williams would win two cents (upper left cell). Since this is a non-zero-sum game, all winnings were paid from a bank. In no case does one player pay the other player anything.

As in the diamond transaction, both players find that one strategy is more profitable no matter what the other player does. Alchian is always better off choosing his strategy 2, and Williams is better off choosing his strategy 1. But when both choose their "better" strategy, both do relatively poorly. They actually do better choosing their "worse" strategies—provided both do it.

The Nash theory suggests the lower left cell (boldface) as the rational outcome. Neither player can do any better by switching strategy unilaterally. In a prisoner's dilemma, the equilibrium point strategy is called *defection.* A player is always better off defecting, no matter what the other does.

But look at the upper right cell. Each player is half a cent better off than he would be in the equilibrium point outcome. The other strategy of a prisoner's dilemma, which leads to the best collective outcome, is called *cooperation.* In the diamond transaction, cheating is defection, and complying with the deal is cooperation.

In the RAND experiment, Alchian and Williams played the game one hundred times in succession. There was no evidence of any instinctive preference for the Nash equilibrium—if anything, the reverse. Alchian chose his nonequilibrium strategy (cooperation; his strategy 1) sixty-eight times out of a hundred, and Williams picked his nonequilibrium strategy (strategy 2) seventy-eight times.

Flood's 1952 paper reports not only the strategies the two players chose but a log of running comments. The comments reveal a difficult struggle to secure mutual cooperation. Williams recognized that the players ought to cooperate to maximize their winnings. When Alchian didn't cooperate, Williams "punished" him by defecting on the next round. Then Williams went back to cooperating. All in all, Williams played quite reasonably—about as most game theorists would play today, after four decades of research.

Alchian began expecting both players to defect. He reports being puzzled at Williams's initial attempts to cooperate. Later in the experiment, Alchian talks of Williams's unwillingness to "share." It's unclear what he meant by this. Possibly he wasn't satisfied with mutual cooperation because (with this particular payoff table) he won only half a cent to Williams's one cent. The payoffs *are* slanted in favor of Williams. Alchian wanted to defect to even things up. When he did, it caused Williams to defect in retaliation.

The comments were apparently written after each player had decided on a strategy in that particular game, but before the other player's choice was known. Some comments thus refer to the other player's choice on the previous game. For clarity, strategy numbers have been replaced with [C] for "cooperate" and [D] for "defect."

Game	AA	JW	AA's comments	JW's comments
1	D	C	JW will play [D]—sure win. Hence if I play [C]—I lose.	Hope he's bright.
2	D	C	What is he doing?!!	He isn't but maybe he'll wise up.
3	D	D	Trying mixed?	Okay, dope.
4	D	D	Has he settled on [D]?	Okay, dope.
5	C	D	Perverse!	It isn't the best of all possible worlds.
6	D	C	I'm sticking to [D] since he will mix for at least 4 more times.	Oh ho! Guess I'll have to give him another chance.
7	D	C		Cagey, ain't he? Well . . .
8	D	D		In time he could learn, but not in ten moves so:
9	D	D	If I mix occasionally, he will switch—but why will he ever switch from [D]?	
10	D	D	Prediction. He will stick with [D] until I change from [D]. I feel like DuPont.	I can guarantee myself a gain of 5, and guarantee that Player AA breaks

even (at best). On the other hand, with nominal assistance from AA, I can transfer the guarantee of 5 to Player AA and make 10 for myself, too. This means I have control of the game to a large extent, so Player AA had better appreciate this and get on the bandwagon.

With small amounts of money at stake, I would (as above) try (by using [C]) to coax AA into mutually profitable actions. With large amounts at stake I would play [D] until AA displayed some initiative and a willingness to invest in his own future. One play of [C] by AA would change me from [D] to [C], where I would remain until bitten.

On the last play, it would be conservative for me to switch to [D], but I wouldn't do so if the evidence suggested that AA was a nice stable personality

Game	AA	JW	AA's comments	JW's comments
				and not in critical need of just a little extra cash.
11	D	C		Probably learned by now.
12	C	C		I'll be damned! But I'll try again.
13	C	C		That's better.
14	C	C		Ha!
15	C	C		(bliss)
16	D	C		
17	C	D		The stinker.
18	C	D		He's crazy. I'll teach him the hard way.
19	D	D	I'm completely confused. Is he trying to convey information to me?	Let him suffer.
20	D	D		
21	D	C		Maybe he'll be a good boy now.
22	C	C		Always takes time to learn.

23	C	C		Time.
24	C	C		
25	C	C		
26	D	C		
27	C	D		Same old story.
28	D	D	He wants more [C]'s by me than I'm giving.	To hell with him.
29	D	D		
30	D	D		
31	D	C	Some start.	Once again.
32	C	C	JW is bent on sticking to [D]. He will not *share* at all as a price of getting me to stick to [C].	---, he learns slow!
33	C	C		On the beam again.
34	C	C		
35	C	C		
36	C	C		

Game	AA	JW	AA's comments	JW's comments
37	C	C		
38	D	C		
39	C	D		The ---.
40	D	D		
41	D	C		Always try to be virtuous.
42	C	C		Old stuff.
43	C	C		
44	C	C		
45	C	C		
46	C	C		
47	C	C		
48	C	C		
49	D	C	*He will not share.*	
50	C	D		He's a shady character and doesn't realize we are playing a 3rd party, not each other.

51	D	C		
52	C	C		He *requires* great virtue but he doesn't have it himself.
53	C	C		
54	C	C		
55	C	C		
56	C	C		
57	C	C		
58	C	C	He will not share.	
59	C	C	He does not want to *trick me.* He is satisfied. I must teach him to share.	
60	D	C		A shiftless individual —opportunist, knave.
61	C	C		
62	C	C		Goodness me! Friendly!
63	C	C		
64	C	C		
65	C	C		

Game	AA	JW	AA's comments	JW's comments
66	C	C		
67	D	C	He won't share.	
68	C	D	He'll punish for trying!	He can't stand success.
69	D	D		
70	D	D	I'll try once more to share—by taking.	
71	D	C		This is like toilet training a child—you have to be very patient.
72	C	C		
73	C	C		
74	C	C		
75	C	C		
76	C	C		
77	C	C		
78	C	C		
79	C	C		
80	C	C		Well.

81	D	C		
82	C	D		He needs to be taught about that.
83	C	C		
84	C	C		
85	C	C		
86	C	C		
87	C	C		
88	C	C		
89	C	C		
90	C	C		
91	C	C	When will he switch as a last minute grab of [D].[1] Can I beat him to it as late as possible?	
92	C	C		Good.
93	C	C		

1. Alchian's comment reads "a last minute grab of (2)," which appears to be a mistake. Williams's strategy 2 was cooperation. His actions indicate that Alchian was concerned about Williams defecting in the last round(s).

Game	AA	JW	AA's comments	JW's comments
94	C	C		
95	C	C		
96	C	C		
97	C	C		
98	C	C		
99	D	C		
100	D	D		

For all the confusion, mutual cooperation was the most common outcome (sixty of the hundred games). Had Flood and Dresher used a "fair" playoff table, the cooperation rate might have been higher yet.

Flood and Dresher wondered what John Nash would make of this. Mutual defection, the Nash equilibrium, occurred only fourteen times. When they showed their results to Nash, he objected that "the flaw in the experiment as a test of equilibrium point theory is that the experiment really amounts to having players play one large multi-move game. One cannot just as well think of the thing as a sequence of independent games as one can in zero-sum cases. There is too much interaction, which is obvious in the results of the experiment."

This is true enough. However, if you work it out, you find that the Nash equilibrium strategy for the multi-move "supergame" is for both players to defect in each of the hundred trials. They didn't do that.

TUCKER'S ANECDOTE

The odd little game in the Flood-Dresher experiment intrigued the RAND community. I asked Flood if he realized the importance of the prisoner's dilemma when he and Dresher conceived it. "I must admit," he answered, "that I never foresaw the tremendous impact that this idea would have on science and society, although Dresher and I certainly did think that our result was of considerable importance . . . I suspect that I was more excited about its importance for practical applications than was Mel, and he more expectant about its implications for game theory."

Flood recalls that von Neumann thought the game was provocative, in a general way, as a challenge to Nash equilibrium theory, but he didn't take their informal experiment entirely seriously. Dresher showed the game to another RAND consultant, Albert Tucker. Tucker was a distinguished Princeton mathematician who knew both von Neumann and Nash (Nash was one of Tucker's students at Princeton).

Stanford University's psychology department asked Tucker to give a lecture on game theory in May 1950. The game Dresher had shown him stuck in his mind. He felt it was interesting from a much broader viewpoint than game theory and decided to discuss it. Since the audience of psychologists had little background in game theory, Tucker decided he needed to present the game as part of a story. Tucker devised a now-well-known dilemma tale and coined the name, "prisoner's dilemma."

These cosmetic services should not be underestimated. Flood's 1952 RAND memorandum was not widely read, and it does not herald the "discovery" of a new dilemma in any case. Flood and Dresher present the prisoner's dilemma experiment as a psychological study. The game was constructed for the experiment, and there is no indication in the paper that similar games might be important in the real world (though Flood and Dresher well appreciated that they were). By posing the game as a dilemma of choice, and by propagating the story in the scientific community, Tucker aided materially in the subsequent evolution of research on social dilemmas.

Tucker sketched the dilemma in a letter to Dresher, calling it a " 'dressed up' version of [a] game like one you showed me." Tucker's capsule description goes,

> *Two men, charged with a joint violation of law, are held separately by the police. Each is told that*
> *(1) if one confesses and the other does not, the former will be given a reward . . . and the latter will be fined . . .*
> *(2) if both confess, each will be fined . . .*

At the same time, each has good reason to believe that (3) if neither confesses, both will go clear.

Over the years, this story has improved in the retelling and now almost always concerns prison terms. (Plea bargains over prison terms are more realistic than cash prizes for confessing!) A typical contemporary version of the story goes like this:

Two members of a criminal gang are arrested and imprisoned. Each prisoner is in solitary confinement with no means of speaking to or exchanging messages with the other. The police admit they don't have enough evidence to convict the pair on the principal charge. They plan to sentence both to a year in prison on a lesser charge. Simultaneously, the police offer each prisoner a Faustian bargain. If he testifies against his partner, he will go free while the partner will get three years in prison on the main charge. Oh, yes, there is a catch . . . If *both* prisoners testify against each other, both will be sentenced to two years in jail.

The prisoners are given a little time to think this over, but in no case may either learn what the other has decided until he has irrevocably made his decision. Each is informed that the other prisoner is being offered the very same deal. Each prisoner is concerned only with his own welfare—with minimizing his own prison sentence.

	B refuses deal	**B turns state's evidence**
A refuses deal	1 year, 1 year	3 years, 0 years
A turns state's evidence	0 years, 3 years	2 years, 2 years

The prisoners can reason as follows: "Suppose I testify and the other prisoner doesn't. Then I get off scot-free (rather than spending a year in jail). Suppose I testify and the other prisoner does too. Then I get two years (rather than three). Either way I'm better off turning state's evidence. *Testifying takes a year off my sentence, no matter what the other guy does.*"

The trouble is, the other prisoner can and will come to the very same conclusion. If both parties are rational, both will testify and both will get two years in jail. If only they had both refused to testify, they would have got just a year each!

Tucker's story was not intended to be a realistic picture of criminology. It's interesting, then, that some corrections experts' misgivings about plea bargaining complain of real-life prisoner's dilemmas. It takes an overwhelmingly strong case to secure a death penalty. Not only must the evidence show that the suspect committed the murder but it must also indicate that the crime was committed in "cold blood." In practice, death penalty (as opposed to life imprisonment) convictions often hinge on the testimony of an accomplice who was in on the plans for the crime. In connection with the death sentence of child murderer Robert Alton Harris, the *Los Angeles Times* wrote (January 29, 1990):

> *In robberies where murders occur, for example, there is often more than one criminal involved and thus more than one person who may be eligible for the death penalty. But for a prosecutor, "what's important is that you score one touchdown," in the form of a death sentence, said Franklin R. Zimring, UC Berkeley law professor and a capital punishment expert.*
>
> *Frequently, a race ensues in which the robbers try to be the first to point the finger at an accomplice and make a deal with the prosecutor to testify in return for leniency.*
>
> *Sometimes, Zimring said, it never becomes clear whether the person who got leniency or the person on trial for his life actually pulled the trigger.*

The one problem with dramatizing the prisoner's dilemma in a story is that it brings in emotional factors that are irrelevant. You might well feel that, were you ever in a situation like either of the above, you couldn't take unfair advantage of the other person. It would be against your moral code; you would feel awful afterward.

All right, think of it as a friendly "game" with no moral implications. No one's going to be upset with you for "cheating" or "squealing"—in fact, let's not use any judgmental words. Imagine a gambling casino that offers a prisoner's dilemma table. Two customers at a time, you vie for cash prizes. Each player decides what to do and indicates the choice by flipping a concealed switch when the croupier says to do so. The two players can't confer with each other about their choices, and house rules forbid any pair from playing the game more than once. It's not morally wrong to try to win as much as possible, any more than it's wrong to play your best at poker or blackjack or any other game. "Good sportsmanship" *demands* that you play to win.

The payoffs (in dollars, francs, casino chips, or any betting unit you like) are:

	B cooperates	B defects
A cooperates	2, 2	0, 3
A defects	3, 0	**1, 1**

The question is, what strategy works best (for you and you alone) when your partner is also looking out for his own interests? Whatever you answer, you must justify it as a means to improve your winnings. It is beside the point to cooperate "because it's the right thing to do."

As before, you're better off defecting no matter what the other player does. You win $3.00 (rather than $2.00) if the other guy cooperates. You win $1.00 (rather than 0) if he defects. It can't get any simpler than that. The casino pays you $1.00 to defect!

If both players are logical and realize this, they will both defect and win $1.00 each. Had they been less "logical" they could have cooperated and won twice as much.

These payoffs are simpler than in the RAND experiment. For a game to be a prisoner's dilemma, it is necessary only that the payoffs are *ranked* in a certain way. In general a prisoner's dilemma takes this form: there is a *reward* payoff (the $2.00 above) for mutual cooperation, which both desire more than the *punishment* payoff ($1.00 above) both receive for not cooperating. But both covet the *temptation* payoff ($3.00 above), the highly desirable outcome of a single defection, even more than the reward. Both fear being the one who doesn't defect and getting stuck with the *sucker* payoff (0 above).

When payoffs are stated in numerical units like dollars, it is usually required that the reward payoff be greater than the *average* of the temptation and sucker payoffs. The full force of the prisoner's dilemma requires that the common good be served by mutual cooperation—hence the peculiar twist that "logical" players cut their own necks by defecting. In the table above, the players win a total of $3 + 0 = $3 when one cooperates and the other defects—an average of $1.50 each, you might say. When both cooperate, each wins $2.00.

Were the average of the temptation and sucker payoffs greater, the players might elect to trade off defecting unilaterally in repeated play and thereby win more money than they would by cooperating. In a true prisoner's dilemma, this is impossible.

I have now described several instances of the prisoner's dilemma, each ending with an odd contradiction. No matter what course you take, you end up wondering if you have chosen correctly. How should one act in a prisoner's dilemma?

In the main this is still an unanswered and probably unanswerable question. Game theorists R. Duncan Luce and Howard Raiffa, who gave the prisoner's dilemma great emphasis in their 1957 book *Games and Decisions,* wrote, "The hopelessness that one feels in such a game as this cannot be overcome by a play on the words 'rational' and 'irrational'; it is inherent in the situation."

COMMON SENSE

The prisoner's dilemma is difficult because it defies commonsense reasoning. Let's see why.

The commonsense argument for defecting goes like this: "A prisoner's dilemma is a simultaneous choice. There is no way that your choice can affect the other player's choice. So the situation is simple. No matter what the other player does, you're better off by defecting. That means you should defect."

The pro-defection crowd even has a rebuttal for the first argument the pro-cooperation camp is likely to bring up, namely, what if everyone reasoned that way? "You say it would be too bad if both players defected when they could have cooperated? Wrong! Remember, the choices don't influence each other. If the other player defects, he defects. My choice had nothing to do with it. Whenever mutual defection occurs, I think how lucky it is that I defected. Had I cooperated, I would have got stuck with the sucker payoff."

The commonsense argument for cooperation is this: "The two players' situations are identical. It is unrealistic for one to expect to take advantage of the other by defecting. Assuming that the players are both rational, they should decide on the *same* strategy. The two realistic outcomes are mutual cooperation and mutual defection. Both prefer the cooperative outcome, so that's what they should do: cooperate."

A little thought shows that this argument doesn't quite hold water.

In a real-life situation, there's no guarantee that both players will make the same choice. Some prisoners rat on their partners and some don't; some people cheat on illicit transactions and others stick to the deal. So as a practical matter, you have to assume that all four outcomes are possible.

The more interesting part of the argument applies to the perfectly rational players postulated by game theory. Presumably there is only one "rational" way to act in a prisoner's dilemma. Therefore only the two mutual outcomes (both cooperate or both defect) are possible in a dilemma played out between these rational players. The argument goes astray only in assuming that the players get to *choose* which of these outcomes is the rational one.

Suppose an eccentric millionaire picks two competent mathematicians on opposite sides of the globe and offers to pay each of them a million dollars times the value of the millionth digit in the decimal expansion of pi. The digit in question could be any of the ten digits and thus the value of the prize could range from zero to $9 million. Logic demands that the mathematicians, working independently, will come up with the same value for the millionth digit of pi. The fact that each would most *like* the digit to be nine does not matter.

Likewise, the fact that perfectly rational players might prefer that their logic lead both to cooperate is beside the point. The question is, what does logic compel them to do?

Another argument for cooperation goes like this: "The best outcome all around is mutual cooperation. The total payout is $6.00, versus $5.00 when players choose different strategies and $2.00 when both defect. So you want to encourage mutual cooperation, and the only way to do that is to cooperate. Even if you get hurt this time, cooperation is the best policy in the long run."

This argument turns out to be quite valid, but not in the present situation. Were a prisoner's dilemma repeated over and over with the same partner, there would be a "long run" to worry about, and the case for cooperation would be much stronger (as we will see). But right now, we're looking at a one-time prisoner's dilemma. You make your choice, and that's that. You might as well get all you can.

In a true, one-time-only prisoner's dilemma, it is as hard to justify cooperation as it is to accept mutual defection as the logical outcome. Therein lies the paradox.

Both Flood and Dresher say they initially hoped that someone at RAND would "resolve" the prisoner's dilemma. They expected Nash, von Neumann, or someone to mull over the problem and come up with a new and better theory of non-zero-sum games. The theory would address the conflict between individual and collective rationality typified by the prisoner's dilemma. Possibly it would show, somehow, that cooperation is rational after all.

The solution never came. Flood and Dresher now believe that the prisoner's dilemma will never be "solved," and nearly all game theorists agree with them. The prisoner's dilemma remains a negative result—a demonstration of what's *wrong* with theory, and indeed, the world.

PRISONER'S DILEMMAS IN LITERATURE

All this would be of only academic importance if the prisoner's dilemma was some hothouse rarity of game theory. Of course, it's not. The prisoner's dilemma is a paradox we all have to live with.

Discovering the prisoner's dilemma is something like discovering air. It has always been with us, and people have always noticed it—more or less.

Ethical prescriptions motivated by prisoner's dilemma-like conflicts are common enough. The Gospel of Matthew (c. 70–80 A.D.) attributes to Jesus the "golden rule": "In everything, do to others what you would have them do to you." (Matt. 7:12). In basic form, this rule was old even then. Earlier versions appear in the writings of Seneca (4 B.C.–65 A.D.), Hillel (fl. 30 B.C.–9 A.D.), Aristotle (384–322 B.C.), Plato (427?–347 B.C.), and Confucius (551–479 B.C.)—and it may not have been original with any of them. It is probably not straining interpretation too much to say that the golden rule addresses prisoner's dilemma-like conflicts. People normally look out for their own interests. There would be no need for the rule unless it sometimes results in the unpopular prescription that people overlook apparent self-interest in order to achieve a mutual benefit possible only when people cooperate.

Similar advice appears in Immanuel Kant's *Critique of Practical Reason* (1788) under the name of the "categorical imperative." Kant concludes that ethical behavior is that which can be universalized. In other words, always ask yourself: What if everyone did that?

None of these prescriptions qualifies as a precocious discovery of the prisoner's dilemma, but the tradition of these philosophers and religious figures is very much part of most people's gut reaction that cooperation is right and defection is wrong.

Close passes to the prisoner's dilemma turn up in Thomas Hobbes's *Leviathan* (1651). In Hobbes's time, monarchs cited divine right to rule. Those of less exalted standing had to accept their status because God decreed it. Hobbes's point in *Leviathan* was that government served a real social function and could be justified, even to the powerless, without theology. Law and order benefit everyone, not just those lucky enough to be dispensing the laws, by preventing defection (to put it in our terms, not Hobbes's). In a lawless society, Hobbes argues, every man is at war with all others, and no one is safe from exploitation. A farmer's crops may be stolen, and thus he has little incentive to plant them in the first place. The society's members are better off surrendering their right to pillage (defect) in return for the security of not being victimized (a reward payoff).

Much closer approaches to the prisoner's dilemma occur in literature. A perceptive discussion of a prisoner's dilemma (closely following Tucker's anecdote!) occurs in Edgar Allan Poe's "The Mystery of Marie Rogêt." Poe's detective C. Auguste Dupin speaks of an offer of reward and immunity to the first member of a criminal gang to confess: "Each one of a gang, so placed, is not so much greedy of reward, or anxious for escape, as *fearful of betrayal.* He betrays eagerly and early that *he may not himself be betrayed."* Poe's story was a thinly fictionalized commentary on a sensational 1842 New York murder. That the real-life reward was never accepted was indication that the murder was not the act of a gang, Poe felt.

Giacomo Puccini's opera *Tosca* (1900) revolves around a clear-cut prisoner's dilemma. The plot comes from an 1887 play by Victorien Sardou. Corrupt police chief Scarpia has condemned Tosca's lover Cavaradossi to death. Scarpia lusts after Tosca and offers her a deal. If Tosca will make love with him, he will instruct the firing squad to use blank cartridges and spare Cavaradossi.[2] Tosca consents. As much as she despises Scarpia, it would be worth granting him her favors to save her beloved.

Should Tosca go ahead with the deal? The two parts of the bargain are effectively simultaneous. Tosca does not have sex with Scarpia until he has given irrevocable orders to use blanks (or real bullets).

2. Puccini favored game-theoretic plots. The heroine of *Girl of the Golden West* bets her virtue against her lover's life in a poker game.

The story ends in mutual defection—the most operatic outcome of a prisoner's dilemma. Tosca betrays Scarpia by stabbing him as they embrace. The tragedy is that Scarpia defected, also. The firing squad uses real bullets, and Cavaradossi is dead. Tosca leaps off a parapet as the police arrive to arrest her for murder.

There is no indication that von Neumann was specifically aware of the game now called the prisoner's dilemma when he wrote *Theory of Games and Economic Behavior.* Even so, von Neumann and Morgenstern were thinking along these lines. The authors write:

> *Imagine that we have discovered a set of rules for all participants—to be termed "optimal" or "rational"—each of which is indeed optimal provided that the other participants conform. Then the question remains as to what will happen if some of the participants do not conform. If that should turn out to be advantageous for them—and, quite particularly, disadvantageous to the conformists—then the above "solution" would seem very questionable. . . . In whatever way we formulate the guiding principles and objective justification of "rational behavior," provisos will have to be made for every possible conduct of "the others."*

The prisoner's dilemma has thus been "discovered," commented upon, and forgotten many times, usually without the realization that it is a universal problem. Political scientist Robert Axelrod told me: "I think it's a little like understanding the role of temperature and heat in physics. You can say, 'It's a hot day,' or 'This glass of water contains more heat than a smaller glass of water which is at the same temperature,' but you can't formulate things very clearly unless you've got the difference between heat and temperature straight. With a prisoner's dilemma, you can say there's a conflict between individual and group interests, but you can't really get very far without the framework of game theory." If so, it is not so remarkable that the RAND group came upon the dilemma six years after the publication of *Theory of Games and Economic Behavior.*

FREE RIDER

It does not take much to create a prisoner's dilemma. The main ingredient is a temptation to better one's own interests in a way that would be ruinous if *everyone* did it. That ingredient, regrettably, is in ample supply. For this reason some have seen in the prisoner's dilemma the fundamental problem of society—the problem of "evil," if you will. The tragedies of history are not the natural disasters but the man-made ones, the consequences of individuals or groups taking actions contrary to the common good.

The most common type of prisoner's dilemma in everyday life is the "free rider dilemma." This is a prisoner's dilemma with many, rather than just two, players.

The name refers to the dilemma confronting public transit riders. It's late at night, and there's no one in the subway station. Why not just hop over the turnstiles and save yourself the fare? But remember, if *everyone* hopped the turnstiles, the subway system would go broke, and no one would be able to get anywhere.

It is the easiest thing in the world to rationalize hopping the turnstiles. What's the chance that your lost fare will bankrupt the subway system? Virtually zero. The trains run whether the cars are empty or full. In no way does an extra passenger increase the system's operating expenses. Etc., etc., etc.—but if everybody thinks this way . . .

There's also the familiar moral dilemma about leaving a note on the windshield of a parked car you've dented. You're certain that no one has seen the accident. It's an expensive car, so that little dent will cost a lot of money to fix—and you can safely assume the owner has insurance. Do you leave a note with your name and address?

By *not* leaving the note, you save yourself the cost of repair. When people don't leave notes, the cost of repair is shifted from the people who caused the damage to the insurance companies. If you think that's too bad for the insurance companies, think again! The insurance companies are only too glad to handle the extra business. They set their rates accordingly. Insurance companies have to pay their expenses and earn a profit on top of that. It might work out that for every $1,000 they pay out in claims, they have to charge $1,500 in premiums. Consequently, by not leaving the note, you might save yourself $1,000 *but* cost the public $1,500 through higher insurance rates.

In New Zealand, newspaper boxes operate on the honor system. Readers are supposed to drop a coin in the payment box, but nothing physically prevents them from taking a newspaper without paying. Evidently, few readers steal, recognizing the consequences of mass defection. In the United States, unlocked newspaper boxes would be unthinkable!

On the other hand, U.S. public television operates on the honor system. Anyone can watch PBS without donating money, but if everyone did that, there wouldn't be public television. In general, a free rider dilemma occurs any time payment for a product or service is on the honor system or where payment, while technically mandatory, is difficult to enforce (people sneaking into ballparks; people underreporting income on their tax forms).

The agonizing decision of whether to meet the demands of terrorists or kidnappers is also a free-rider dilemma. Normally one prefers to pay the ransom as long as the hostage is returned safely. But paying a ransom encourages other kidnappers and makes it likely that others will be taken hostage in the future. If no one ever paid a ransom, there would be no kidnappings. Unlike most of the previous examples, this dilemma cannot be resolved by recourse to conventional ideas of morality. It may be "wrong" to hop subway turnstiles, but few would condemn anyone for paying a ransom to secure the release of an innocent party. Surely the kidnappers, and only the kidnappers, are in the wrong, as that term is usually understood. But what should you do?

The free-rider dilemma is even more hopeless than the two-person prisoner's dilemma. No longer is one cooperating or defecting with a single partner. There are many other parties (millions in the case of an urban transit system or a PBS station). Defectors in a free-rider dilemma can hide in the crowd, so to speak. Human psychology being what it is (variable), there can be no doubt that many people will defect and hop the turnstiles. Since there will always be people who "get away with" not paying, the others are suckers who pay full fare but ride a poorly maintained subway because of the revenue lost to turnstile hopping.

Taxes are one way government avoids free-rider dilemmas. It would be nice if everyone voluntarily contributed money for maintaining roads, running schools and post offices, and the hundreds of other functions of government. But few people would do so knowing that many others would pay nothing. Most people can be convinced that taxes for public works are desirable, provided that everyone pays. Thus the government enforces payment of taxes.

Free-rider defection is one of the usual objections to the socialism of Thomas More's *Utopia* and Karl Marx. If everyone worked hard and goods were shared intelligently, no one would starve or want for basic necessities. At the same time, everyone would be tempted to loaf, knowing that they would still eat and that their personal shirking could not much threaten the collective. If everyone did that, the commune would fail and people would starve.

Arguably, the prisoner's dilemma is behind much political debate. Conservatives and liberals don't really want such different things. Liberals don't enjoy paying taxes, and conservatives don't like seeing the homeless encamped in railroad stations. Why then do reasonable people differ so widely in their politics?

As the term is most commonly used in U.S. politics, a liberal is a "cooperator": someone willing to put himself at risk for exploitation in order to increase the common good. Liberals favor paying taxes that go to help the homeless in the expectation that the homeless will not fritter away such aid but will use it to get on their feet. A liberal may favor cutting back in defense expenditures in the hope that other countries will do the same. By cooperating, a liberal expects to create a society with fewer homeless, or fewer missiles—something that everybody wants, but which will not come about through anyone's unilateral effort.

Conservatives are often "defectors" in that they seek to guarantee themselves the best outcome possible on their efforts alone. Taxes may be squandered, so the safest course is to let people keep as much of their income as possible and decide individually how best to spend it. Enemy nations may exploit a unilateral arms freeze to gain the upper hand. Conservative political positions avoid the "sucker payoffs" of welfare cheats and arms treaty violators.

Social problems are never simple. Many political and military debates are so enmeshed in contingencies and uncertainties that it is possible to argue forever over the details. There is a sense that if only these ancillary issues were resolved, the dilemma would be slain and all would agree on the proper course to take. This is not necessarily so. In many cases the central dilemma is genuine and seemingly irresolvable. To the extent that a real social problem poses a prisoner's dilemma, it will be an agonizing choice even when all the side issues are settled. There will be no "right" answer, and reasonable minds will differ.

NUCLEAR RIVALRY

The prisoner's dilemma can be "oversold." There are seventy-eight distinct two-person, two-strategy games. Every one must occur somewhere in real life. Most of the seventy-eight games have obvious solutions. Game theorists have tended to focus on the prisoner's dilemma precisely because it is a problematic case. Most conflicts are not prisoner's dilemmas, though.

No example of a prisoner's dilemma has been more popular, both in technical articles and the popular press, than a nuclear arms rivalry. This is so much the case that the term "prisoner's dilemma" is sometimes taken to be part of the jargon of nuclear strategy, along with "mutual assured destruction" and "MIRV." The perception that arms races pose prisoner's dilemmas may overshadow the reality, but it has become one of the paradigms of our time.

Flood says he wasn't thinking specifically of the nuclear situation when he and Dresher formulated their game—rather, of Nash equilibriums. Of course, it quickly became apparent that there were parallels, and defense in the nuclear age was the underlying purpose of all RAND's research.

At the time the prisoner's dilemma was discovered, the United States and the Soviets were embarked on an expensive nuclear arms race. Whether this race was properly a prisoner's dilemma depends on the motivations of those in power. These motives are open to debate.

For the sake of simplification, suppose that two rival nations must decide whether to build an arsenal of H-bombs or not. It will take years to build a thermonuclear arsenal, and the work may well be done in secret. Each nation must commit to its choice not knowing what the other has decided (until it's too late).

Each nation prefers to be the stronger, the result if it builds the H-bomb and the other nation doesn't. Conversely, each nation is afraid of being the weaker, the one without the H-bomb.

There's little if anything gained should *both* nations get the H-bomb. Geopolitical power depends on relative military strength. Two H-bomb arsenals more or less cancel each other out. Furthermore, it costs a lot of money to build H-bombs. Both nations are materially poorer than they would have been. Worse yet, once a weapon is built, it tends to get used, eventually. No one's going to sleep as soundly as he used to. The weapons that were built to make each side more secure could have the opposite effect.

Under these conditions, building the bombs can be identified with defection, holding off can be identified with cooperation, and the situation is a prisoner's dilemma. Each side would prefer that no one build the bomb (reward payoff for mutual cooperation) rather than both build it for no net gain of power (punishment payoff for mutual defection). But each side may well elect to build the bomb either out of hope of gaining the upper hand militarily (temptation payoff) or out of fear of being the one without it (sucker payoff).

In 1949, the General Advisory Committee (GAC) to the Atomic Energy Commission advised against developing the hydrogen ("super") bomb. The GAC's report said:

> *We all hope that by one means or another the development of these weapons can be avoided. We are reluctant to see the United States take the initiative in precipitating this development. We are all agreed that it would be wrong at the present moment to commit ourselves to an all-out effort towards its development.*
>
> *In determining not to proceed to develop the Superbomb, we see a unique opportunity of providing by example some limitations on the totality of war, and thus eliminating the fear and raising the hopes of mankind.*

On October 31, 1949, GAC head J. Robert Oppenheimer presented these recommendations to Secretary of State Dean Acheson. Acheson later confided to his chief nuclear adviser, "You know, I listened as carefully as I know how, but I don't understand what Oppie is trying to say. How can you persuade a hostile adversary to disarm 'by example'?" This exchange acutely points up the dilemma.

Preventive war's supporters feared a future nuclear deadlock in which either side could launch a devastating surprise attack with minimal retaliation. Whether this hypothetical state of affairs would be a prisoner's dilemma depends, again, on how the sides rank the potential outcomes. Obviously each side would prefer an outcome where it is not attacked to one where it is. For a prisoner's dilemma to exist, it is further necessary that each side prefer to attack the other (even to peace). This is not necessarily the case—in fact, only for a pair of overtly belligerent nations would it be. Unfortunately, fear can be self-amplifying. By 1950, many in the United States and in the Soviet Union saw the other nation as an implacable foe.

7

1950

The year 1950 was the first to dawn with atomic bombs on both sides of the iron curtain. The very month of the Flood-Dresher experiment, President Harry Truman decided to build the H-bomb. The following months saw a heating of East-West tensions, culminating in a call for a preventive war against the Soviet Union. The initial reactions to the two-sided nuclear dilemma, both official and unofficial, embrace many of the concerns typical of a prisoner's dilemma. The events of the early cold war have become so much a part of the popular idea of what a prisoner's dilemma is that it is worth considering them at some length.

THE SOVIET BOMB

In September 1949, the United States detected indirect but unmistakable evidence of a Soviet atomic bomb. No one had seen a mushroom cloud or felt a shock, even by seismograph. But one of the Air Force's B-29's had detected radioactivity in an air sample taken over Japan. Navy-supplied rainwater samples, collected on ships and at bases worldwide, showed traces of cerium 141 and yttrium 91, products of fission. It was hard to explain these findings in any other way than as fallout from an atomic blast somewhere in central Asia.

Truman called J. Robert Oppenheimer to Washington to ask him if the reports were true. Oppenheimer said they were. The National Security Council debated whether to make the news public. Secretary of Defense Louis Johnson opposed disclosure, fearing panic. He cited Orson Welles's radio broadcast of *War of the Worlds* as an example of what might happen. Secretary of State Dean Acheson favored telling the public, partly on the grounds that the news would be less distressing coming from the President than from the Soviets. Would the Soviets announce it, and when, they wondered. Opinion shifted to Acheson's side when it was learned that Soviet Deputy Minister Andrei Gromyko had scheduled a major speech at the U.N. Truman decided to beat him to the punch.

At 11:02 A.M. on September 23, 1949, the President announced, "I believe that the American people, to the fullest extent consistent with national security, are entitled to be informed of all developments in the atomic energy field. . . . We have evidence that within recent weeks an atomic explosion occurred in the U.S.S.R. Ever since atomic energy was first released by man, the eventual development of this new force by other nations was to be expected. . . . This recent development emphasizes once again . . . the necessity for that truly effective and enforceable international control of atomic energy which this government and the large majority of the members of the U.N. support."

Two days after Truman's announcement, the Soviet Union conceded that it had the atomic bomb "at her disposal."

To the public and much of the military, the Soviet bomb came as a surprise. No one doubted that the Soviets were working on a bomb. Dozens of estimates put the first successful test years in the future, though. In 1945 Leslie Groves had speculated, based on his experience at Los Alamos, that it would take the Soviets fifteen to twenty years to make a bomb. Vannevar Bush had guessed twenty years in 1946.

In retrospect, these estimates are wildly optimistic, and even insulting. Why would it take the Soviets twenty years to duplicate what the United States had done in three? The Soviets had the advantage of knowing the bomb was possible, plus such clues as they could get from fallout—not to mention secrets learned from a successful espionage operation. Chauvinistic U.S. leaders felt that the Soviets were technologically backward: "They can't even manufacture a decent saucepan." (In fairness, such attitudes were encouraged by years of Soviet bluffing. In 1946, the Soviet delegate to the UN boasted that in his country atomic energy was used for peaceful purposes only—like changing the courses of rivers and moving mountains.)

Oddly, Truman was long skeptical of the Soviet bomb's existence—literally for years afterward. In his postpresidential memoir, *Years of Trial and Hope* (1956), Truman wrote: "I am not convinced that the Russians have the bomb. I am not convinced that the Russians have achieved the know-how to put the complicated mechanism together to make the A-bomb work." Truman suspected that the detonation might have been an accident—the explosion of a laboratory—rather than a planned test. It is difficult to square that with the fact that *two* Soviet explosions were detected in October 1951 (one on the third and one on the twenty-second).

Most Americans accepted the bomb and speculated nervously on whether, or how, the Soviets were capable of delivering it to a target. On May Day, 1947, Western observers had identified B-29-like Soviet planes flying in formation over the Kremlin. These had apparently been patterned after American planes that landed in Siberia during the war. The Russians were also known to have copied German submarines, and these might be used to launch atomic missiles. These suppositions inflamed wilder speculation. Ships could enter a harbor, silently deposit a time-delay bomb on the harbor floor, and sail out. The bomb could fit in a truck and be driven to its American target, *Life* magazine warned its readers. The *Bulletin of the Atomic Scientists* moved the minute hand of its clock forward to three minutes to midnight.

Some worried that the Soviet fission bomb would quickly lead to deadlier weapons. According to James R. Shepley and Clay Blair, Jr.'s *The Hydrogen Bomb* (1954), Senator Brien McMahon shouted at AEC representatives: "How do you know that the Russians did not decide from the beginning to shoot for a *hydrogen* bomb? How do you know that the Soviets, now that they have the A-bomb trigger, will not fire off an H-bomb next month?"

THE MAN FROM MARS

The Soviet bomb got many more Americans thinking seriously about preventive war. Truman's science adviser, William Golden, wrote Lewis Strauss shortly after the Soviet bomb blast (letter dated September 25, 1949, filed with Golden's papers at the Truman Library). His words are typical of the mixed feelings many had. Golden keeps preventive war at arm's length, as something only a "Disinterested Man from Mars" might propose:

> *This brings up the matter of immediate use, or threat of use, of our weapons. Let us not delude ourselves, to bring about a true international control agreement with Russia we would have to use them. The consequences would be dreadful indeed, even though I assume that the Russians have so few A bombs* now *that they could do little or no damage to the U.S.A. even if they could put them on target.*
>
> *In theory we should issue an ultimatum and use the bombs vs.*

Russia now: For, from here on, we inevitably lose ground. And this is true no matter at how much greater a rate we produce no matter how much more potent weapons. For once Russia is in a position to put A bombs on our cities, no matter how inefficient those bombs may be and how few in number, she is in a position to do us unspeakable injury. That we can retaliate a hundredfold, or wipe out every Russian, will not repair the damage. So a good, though amoral case can be made by the Disinterested Man from Mars for our shooting at once. Studies along this line, and in more dilute version, should and doubtless will be (or have been) made by the Joint Chiefs. . . . However, we won't do it, of course; no matter what the alternative cost in the long run, the public would never support so far-seeing a bombardment.

Golden recommends that preventive war be kept in mind, however, and that bomber crews be kept in readiness.

UREY'S SPEECH

In New York on January 27, 1950, chemist Harold C. Urey of the University of Chicago gave a dinner speech about the still-hypothetical hydrogen bomb. Urey was not working on a bomb, but he spoke with authority. He had been the first to isolate deuterium ("heavy hydrogen"), which would soon be a key ingredient of the H-bomb. This feat had earned him the Nobel prize in 1934.

Joseph E. Mayer, also of the University of Chicago, was sufficiently convinced that Urey's talk would interest John von Neumann that he sent him an advance copy for comment. Urey posed three related scenarios:

. . . let us assume that the USSR is developing this bomb; and suppose that she should get it first. Then it seems to me that there is nothing in the temperament of the present negotiations between east and west that would lead us to believe that the rulers of the USSR would not reason approximately as follows:

"It is true that the bomb is exceedingly dangerous, and we would not wish to produce so much radioactivity in the world as to endanger ourselves and the people of Russia, but the explosion of a few of these bombs will win us the world. Therefore we will

build these bombs and issue ultimata to the western countries, and the millennium of Communism will be with us immediately. After this the universal government of the USSR will abolish all stocks of bombs and no more will ever be made in the world."

This is a very good argument. In fact, I doubt if any bombs would need to be exploded. The atomic bomb is a very important weapon of war, but hardly decisive, as everybody has emphasized from the beginning. But I wonder if the hydrogen bomb would not be decisive, so that ultimata would be accepted and it would be unnecessary to deliver the bomb. This seems to me to be the situation.

. . . Suppose that we get the hydrogen bomb first. Then what do we do? Do we merely wait until the USSR also has it and we have a stalemate? It is quite out of character for the democracies to deliver ultimata with the philosophy I have ascribed to the Russian government. This is one of the things I cannot answer, and which I merely put before you. . . .

Suppose that two countries have the hydrogen bomb. Is it not believable that sooner or later an incident may occur which would make these bombs be used? This is a question again which I cannot answer definitely. I would say, however, that the probability that a war will start is increased if two groups each believe that they can win the war. This is true regardless of weapons and their magnitude. It is very difficult to get an exact balance of power. This is what we know in physical science as a situation of unstable equilibrium; one like balancing an egg on its end. The slightest push topples the egg in one direction or another.

Urey concluded that "there is no constructive solution to the world's problems except eventually a world government capable of establishing law over the entire surface of the earth." He recommended "any steps of any kind whatever that move in this direction," including "the establishment of an Atlantic Union over the democratic countries of the world . . . and the extension of any such organization to as much of the rest of the world as possible and in as short a time as possible."

THE FUCHS AFFAIR

By coincidence, a shocking bit of news was revealed the same day as Urey's speech. The British embassy informed the United States that physicist Klaus Fuchs had been arrested as a Soviet spy in London.

Though German-born, Fuchs was one of Britain's top physicists. He had been head of Harwell, Britain's secret atomic research center. Fuchs had worked at Los Alamos through the summer of 1946. As leaks go, the situation could hardly have been worse. The publications department at Los Alamos had produced a top-secret *Disclosure of Invention* paper. This was a summary of all the potentially patentable discoveries made at Los Alamos, written to protect the financial interests of the discoverers and their heirs. The paper contained essentially everything then known about atomic weapons—including lines of speculation for an H-bomb (the "super"). It had been written by Klaus Fuchs and John von Neumann.

That Fuchs and von Neumann were asked to write the patent paper testified both to the trust placed in them by the Los Alamos community and to their encyclopedic knowledge of the work there. Nuel Pharr Davis recounts in his *Lawrence & Oppenheimer* (1968):

> *All security-cleared scientists in the country thought at once of the Fuchs-von Neumann patent. To check on how much it contained about the Super, Bethe in New York telephoned Ralph Smith, director of the documents division at Los Alamos.*
>
> *"Is it all there?" Bethe asked.*
>
> *"All," said Smith.*
>
> *"Oh," replied Bethe in a tone Smith found hard to distinguish from the moaning of the long-distance wires.*

Fuchs came of age as Nazism was overtaking Germany. In 1932, at the age of twenty-one, he joined the Communist party, one of the few distinctly anti-Nazi political forces remaining in Germany. After the Nazis blamed the 1933 Reichstag fire on the Communists, it became dangerous to admit party membership. The morning following the blaze, "I took the hammer & sickle from the lapel of my coat where I had carried it," Fuchs stated in his confession. "I was ready to accept the philosophy of the party as right in the coming struggle." Shortly thereafter, Fuchs emigrated to Britain. He was sent to Canada for a time and allowed to return to England, ironically, only after it had been determined that he was not a Nazi.

Fuchs was not a spy by trade. Had he been, he could not have achieved the position of trust he held. He was a fully qualified and talented physicist. He was an expert on isotope diffusion, the process by which radioactive elements are purified for use in the bomb. Fuchs was formally accused of transmitting atomic secrets to the Soviets four times between 1943 and 1947. For this he received only token payment, the largest amount being $400.

As spies must be, Fuchs was pleasant and did not stand out in a crowd. He once lent his car to Richard Feynman so he could visit his wife in the hospital. In *The Legacy of Hiroshima* (1962) Teller wrote of Fuchs, "He was by no means an introvert, but he was a quiet man. I rather liked him. . . . Fuchs was popular at Los Alamos because he was kind, helpful and much interested in the work of others."

History has downplayed the military significance of Fuchs's spying. He may have accelerated the Soviet Union's work on the fission bomb, but that, after all, was an accepted fact when the Fuchs story hit the papers. The H-bomb "secrets" Fuchs passed were of Teller's initial idea, a design that turned out to be unworkable. The ultimately successful Teller-Ulam design for the H-bomb was conceived in 1951. By then Fuchs was behind bars in Britain. Current opinion holds that the greatest boost to the Soviet H-bomb program probably came from analysis of the fallout from U.S. H-bomb tests. From the proportions of isotopes in the fallout, a physicist of the stature of Sakharov could have deduced the high compression of the deuterium and from that, the basic design of the bomb.

At the time, however, the Fuchs news was devastating both to hawks and doves. Things moved quickly after Fuchs's arrest. Four days later, on January 31, Truman initiated the accelerated hydrogen bomb program. The official statement soft-pedaled the decision: ". . . I have directed the Atomic Energy Commission to continue its work on all forms of atomic weapons, including the so-called hydrogen or super bomb."

Reactions were mixed. A concerned woman in Augusta, New Jersey, wrote Truman: "Please do not make a hasty decision on that horrible bomb. Isn't the atomic one bad enough?" But U.S. Commissioner John J. McCloy declared, "I am glad of President Truman's decision. If there were an oxygen bomb that would be bigger than the H-bomb, I would build it."

Fuchs's short trial took place at London's Old Bailey on March 1, 1950. The press and a number of distinguished spectators packed the courtroom. The Duchess of Kent appeared in the visitors' gallery with a rose corsage. Scarcely noticed was Sir Percy Sillitoe, the head of MI-5, the British counterespionage agency that had failed to catch Fuchs.

Bewigged and scarlet-gowned, Lord Goddard heard the evidence, occasionally taking a pinch of snuff from a silver box. The press painted Fuchs in the terms reserved for spies: "slight, sallow, and weak-chinned, in [a] well-pressed light-brown suit matching his thinning hair," said *Newsweek*. He pleaded guilty in a high, tinny voice.

Fuchs's attorney offered little in his defense except to say that Fuchs had never pretended to be anything but a Communist, and that there was no evidence to convict him beyond his own confession. Fuchs mentioned something about "controlled schizophrenia." He claimed that half his mind was Communist and the other half loyal to Britain. Lord Goddard didn't buy that: "I cannot understand all this metaphysical talk, and I don't know that I should."

Goddard commented that Fuchs's crime bordered on high treason, a capital crime. He concluded however that it was not high treason and therefore was subject to a maximum penalty of fourteen years' imprisonment. He imposed this maximum sentence. Fuchs thanked all concerned for a fair trial.

Fuchs served nine years of the sentence. In 1959 he was turned over to East German authorities. East Germany greeted Fuchs as a returning hero and quickly offered him a post at the Institute for Nuclear Physics near Dresden. Fuchs served there with distinction for two decades after his release, retiring in 1979. He died in 1988.

The advance copy of Urey's speech reached von Neumann late as he had been out of town. Von Neumann wrote back to Mayer on February 3, when the situation was much changed. Fuchs had been arrested, and the United States was building an H-bomb. "The apparent hesitation of the commission and/or of its advisors on the subject of the hydrogen bomb is something where my feelings are still stronger, and my technical inhibitions very considerably less," von Neumann wrote. "I think that there should never have been any hesitation on this subject."

On what actually should be done once we had the H-bomb, von

Neumann is noncommittal. "I agree with you completely that one should distinguish between Urey's general analysis of the situation and the particular political solution that he proposes. It is not necessary to discuss the latter at this point. I certainly agree in all essentials with former."

THE KOREAN WAR

On June 25, 1950, the growing pessimism was confirmed by North Korea's invasion of South Korea. The Korean War caused the American public and leadership to rethink their views of Soviet intentions. In the years immediately after World War II, most American leaders had seen the Soviets as too weak to risk a war. Nearly everyone felt that the Soviets desired peace—if nothing else, to give them a chance to rearm. Secretary of Defense Johnson told the U.S. Senate Committee on Appropriations (1950):

> *The very fact of this aggression . . . constitute[s] undeniable proof that the forces of international communism possess not only the willingness, but also the intention, of attacking and invading any free nation within their reach at any time they think they can get away with it. The real significance of the North Korean aggression lies in this evidence that, even at the resultant risk of starting a third world war, communism is willing to resort to armed aggression, whenever it believes it can win.*

Senator Lloyd Bentsen, Jr., urged the use of the atom bomb in Korea—or Manchuria. Even Truman spoke of the possibility of using the bomb in Korea. These and other words from ranking officials helped legitimize the idea that the United States might use atomic weapons first, even in a world in which it was not the sole nuclear power.

THE NATURE OF TECHNICAL SURPRISE

It is claimed, wrongly, that nothing changes in human affairs. The speed and scale of devastation created by atomic weapons completely changed the nature of war. Von Neumann's was the first generation in which game theory or something like it could apply so broadly to warfare.

The advantage of a surprise attack was appreciated in the most ancient writings on war. The bomb raised the possibility of a surprise attack winning not just a battle but the war. In a 1955 study, *Defense in Atomic War,* von Neumann wrote:

> *In the past . . . if the enemy came out with a particularly brilliant new trick, then you just had to take your losses until you had developed the countermeasures, which may have taken weeks or months. The period of one month is probably reasonable for a very brilliantly performed counter-counter-move. This duration is now much too long, and the losses you may have to take during this period may be quite decisive. . . . The difficulty with atomic weapons, and especially with missile-carried atomic weapons, will be that they can decide a war, and do a good deal more in terms of destruction, in less than a month or two weeks. Consequently, the nature of technical surprise will be different from what it was before. It will not be sufficient to know the enemy has only fifty possible tricks and that you can counter every one of them, but you must also invent some system of being able to counter them practically at the instant they occur.*

It became possible to conceive that a surprise atomic attack could seal the fate of the nation under attack even before the nation's leaders were aware that war had begun. Against such possibilities (however realistic or unrealistic they may have been in 1950) conventional statesmanship and military strategy are of little use. Statesmanship is normally a matter of making appropriate responses to threats. A nation demonstrates willingness and ability to defend against encroachment in the hope that, most of the time, this defense ability would be accepted and render fighting unnecessary. The bomb seemingly swept that aside. A nation was always vulnerable to a devastating attack that might come with no warning, and there was no sure defense except to strike first. With life-or-death choices, made in ignorance of simultaneous decisions, war imitates game theory.

By 1950 von Neumann was known to associates as a resolute supporter of preventive war. Written in the midst of the cold war (1957), a generally flattering obituary of von Neumann for *Life* magazine by Clay Blair, Jr., relates:

After the Axis had been destroyed, Von Neumann urged that the U.S. immediately build even more powerful atomic weapons and use them before the Soviets could develop nuclear weapons of their own. It was not an emotional crusade, Von Neumann, like others, had coldly reasoned that the world had grown too small to permit nations to conduct their affairs independently of one another. He held that world government was inevitable—and the sooner the better. But he also believed it could never be established while Soviet Communism dominated half of the globe. A famous Von Neumann observation at that time: "With the Russians it is not a question of whether but of when." A hard-boiled strategist, he was one of the few scientists to advocate preventive war, and in 1950 he was remarking, "If you say why not bomb them tomorrow, I say why not today? If you say today at 5 o'clock, I say why not one o'clock?"

It is hazardous to guess another person's inner motivations. Paul Halmos attributed von Neumann's support of preventive war to his strong dislike of communism. Certainly von Neumann's childhood recollections of the Kun regime were reason for a distrust of communism. As a Hungarian, von Neumann must have been aware of Russia as a historical enemy as well. During his testimony at the Oppenheimer hearings, von Neumann himself said, "I think you will find, generally speaking, among Hungarians an emotional fear and dislike of Russia."

Von Neumann was enamored with Thucydides's *History of the Peloponnesian War.* The latter is often considered a preventive war. Thucydides writes, "What made the war inevitable was the growth of Athenian power and the fear which this caused in Sparta." A passage von Neumann could quote word for word—and did during the time he was advocating preventive war—was the coldly rational advice of the militarily strong Athenians to the weaker Melians:

We recommend that you should try to get what it is possible for you to get, taking into consideration what we both really do think; since you know as well as we do that, when these matters are discussed by practical people, the standard of justice depends on the equality of power to compel and that in fact the strong do what they have the power to do and the weak accept what they have to accept.

This has the flavor of game theory in its suggestion that the rational outcome of conflict may be neither fair nor unanimously desired. The Melians' only consolation is that they could not have done any better, given the greater strength of the Athenians.

It's unlikely that von Neumann—or anyone else, circa 1950—explicitly thought of the U.S.-Soviet conflict as a prisoner's dilemma. If von Neumann did picture U.S.-Soviet relations as a game, it is more plausible that he saw it a zero-sum game. So it would be if the United States and Soviets are pictured as implacable enemies. One of the very few military leaders of the time aware of the prisoner's dilemma was General Andrew Goodpaster, a friend of Merrill Flood's. In 1950, General Goodpaster spent several weeks at RAND, and Flood told him about the prisoner's dilemma. Flood thought that the recognition that military decisions might pose a prisoner's dilemma could be useful. If nothing else, the prisoner's dilemma cautioned that certain decisions were not as pat as they might appear. Flood told me, "Our chats were interesting but never led to any serious action on Andy's part, so far as I know, that were inspired by our discussions of game theory ideas—including dilemma game ideas—a disappointment from my standpoint, especially since Andy was not only a distinguished general but also educated at Princeton, with a doctorate in political science. I always felt that if Andy could not or did not do this nobody could or would."

Nevertheless, many statements of those in power express concerns characteristic of a prisoner's dilemma. That the Soviets might defect—launch a surprise atomic attack on the United States—was the implication of Winston Churchill's comment (1948) when he was asked what would happen when the Soviets got the bomb. Churchill said, "You can judge for yourself what will happen then by what is happening now. If these things are done in the green wood, what will be done in the dry?" General Omar Bradley was quoted as saying, "If Russia had the atomic bomb I don't believe they would hesitate to use it on us."

It was further realized that a nuclear war would be a virtually simultaneous exchange. As early as the 1945 hearings on Pearl Harbor, Senator Brien McMahon said, "If there is ever an atomic Pearl Harbor, there won't be a coroner's jury of statesmen left to talk about it." Major General Robert M. Webster stated, "I believe any attack against us will be attempted as a complete and utter surprise."

AGGRESSORS FOR PEACE

It was 3 A.M. of a deep summer night in the Midwest. The Secretary of the Navy couldn't sleep. Francis P. Matthews was home in Omaha on a vacation from his official duties in Washington. He didn't like the speech a Navy writer had prepared for him to deliver in Boston. Matthews started writing his own speech in the middle of the night and finished it by ten the next morning.

The speech Matthews had written called for the United States to wage war against an unnamed enemy that was pretty obviously the Soviet Union, using unspecified weapons that would presumably include atomic bombs. In accordance with protocol, Matthews made two copies of the speech and mailed them to his office in Washington for clearance through official channels. Matthews was well aware of the regulations requiring clearance of official speeches, for he himself had authored those regulations.

The speech should have gone to Stephen T. Early, Under Secretary of Defense, for content approval. But Early didn't see it. Matthews's office sent the speech directly to the Public Information Office of the Defense Department. Here stories differ. Either the staffers in the Public Information Office assumed that Early had already cleared the speech, or they weren't aware that the speech needed approval—wasn't it just a rewrite of the discarded speech, for a routine ceremony? The Public Information Office mimeographed the speech and distributed it to the press.

Matthews's notorious Boston speech was not quite his first public statement favoring preventive war. Two days earlier (August 23, 1950), Matthews had given an impromptu noontime talk before the Rotary Club at Omaha's Hotel Fontenelle. His remarks were covered on the front page of the *Omaha World-Herald*. It quoted him as saying, "I would not limit the means we would use to get peace. We should be prepared to use military power if necessary . . . Communist nations are going to aggress when and where they please . . . If we sit idly by, we may suffer."

None of the fellow Rotarians who heard these words objected much to them or considered them controversial. Emboldened by this uncritical reaction, Matthews composed his Boston speech in the wee hours of the next morning.

August 25, 1950, was celebrated as the one hundred and fiftieth anniversary of the Boston Naval Shipyard. Boston Mayor John B. Hynes unveiled a monument to the war dead. Sailors donned 1800-vintage uniforms and scrambled up the rigging of *Old Ironsides.* That evening Matthews and Secretary of Labor Maurice J. Tobin gave speeches. This time the national press was listening.

Matthews's speech started about the way anyone would have imagined such a speech to start. He said nice things about the Navy, the Marines, the shipyard, and America. He asked his listeners, what if we had lost the Revolutionary War? He concluded that there would be no Fourth of July, no Liberty Bell, no Washington Monument, and no Lincoln Memorial. Matthews segued from this to the less obvious assertion that America was the custodian of the Holy Grail and the repository of the Ark of the Covenant. The Holy Grail, Matthews said, was the inspiration for the Declaration of Independence, and the Magna Charta too.

Matthews saved his punch for the last few paragraphs. He asked the nation to consider a war of aggression for peace. He said:

> *. . . A true democracy ordinarily does not seek international accord through result to violence. For one hundred and sixty-three years, the United States has settled its international differences through peaceful negotiation. Never have we drawn the sword first unless first attacked and so compelled to fight in self-defense. It is possible that we shall be forced to alter that pacific policy. . . .*
>
> *We should first get ready to ward off any possible attack and, reversing the traditional attitude of a Democracy, we should boldly proclaim our undeniable objective to be a world at peace. To have peace we should be willing, and declare our intention to pay, any price, even the price of instituting a war to compel cooperation for peace.*
>
> *Only the forces who do not want peace would oppose our efforts to transform the hostile nations embroiled in the present international conflicts into a tranquil world. They would brand our program as imperialist aggression. We could accept that slander with complacency, for in the implementation of a strong, affirmative, peace-seeking policy, though it cast us in a character new to a true democracy—an initiator of a war of aggression—it would*

win for us a proud and popular title. We would become the first Aggressors For Peace.

After the speech, there were fireworks—literally and figuratively. Within hours, the world was asking: Who is Francis Matthews?

FRANCIS MATTHEWS

The same question had been asked little more than a year before, when Truman nominated Matthews as Secretary of the Navy. Matthews was a successful Omaha attorney who, at age sixty-two, had never held a major public office.

Matthews had been raised in humble circumstances in the Midwest and Mississippi. His father ran a country store in Albion, Nebraska. After his father's death, his mother used the insurance proceeds to buy a farm. Matthews built up a law practice and had a number of business interests, including part ownership of an Omaha radio station, a loan company, and building supply firms.

Matthews was a joiner. He was Supreme Knight of the Knights of Columbus, a member of the Rotary Club, a major benefactor of Father Flanagan's Boys Town, and a director of the Boy Scouts, Girl Scouts, *and* Camp Fire Girls. He became active in the Democratic party of Nebraska. If there was one distinguishing quality of his personality, it was his deep religious convictions. Matthews was such a devout Catholic that he built a chapel in his home and prayed there daily. In 1944 he had an audience with Pope Pius XII, who designated him Secret Papal Chamberlain with Cape and Sword. That honor entitled him to serve on the Vatican staff should he ever desire to do so.

In 1946, Truman appointed Matthews to the President's Committee on Civil Rights. The same year Matthews, a member of the national board of the United States Chamber of Commerce, chaired a committee on communism. This group produced and distributed pamphlets with titles such as *Communist Infiltration in the United States, Its Nature and How to Combat It.* One publication charged that "forces within the State Department" were supporting the Chinese communists and claimed that "a real service to the community could be rendered if the secret stories of Yalta and Tehran could be made public."

Despite the fact that this pamphlet was critical of both the Roosevelt and Truman administrations, it did not appear to lessen Matthews's influence with the White House. During the 1948 Democratic convention, Matthews quelled an "abandon Truman" move by Nebraska delegates. The twelve electoral votes were greatly needed, and Truman did not forget the favor. During the campaign, Matthews also became friendly with Louis Johnson, the Truman fund-raiser who would be appointed Defense Secretary.

Matthews was Truman's third choice for the job of Secretary of Navy. The President had preferred Jonathan Daniels, son of Woodrow Wilson's Secretary of the Navy. Daniels wasn't interested. A second choice, Judge Robert Quinn of Rhode Island, was vetoed by Secretary of Defense Johnson. Matthews confessed to surprise at being chosen for the post. "I didn't lift a finger to get it," he told the press. Matthews had never been in the armed forces. During the war, he had toured the British Isles, supervising Catholic relief work and supposedly investigating the religious needs of fighting men. He was no less a landlubber than his Nebraska residence implied. He joked that his seamanship was limited to use of a rowboat at a summer home.

Others were surprised, too. The *St. Louis Post-Dispatch* wrote (May 20, 1949), "It passes understanding that a man associated with unfair criticism of the Administration . . . should be invited to become a part of it."

By most accounts, Matthews was a likable guy who was not very good at his job as Secretary of the Navy. Besides knowing almost nothing about the Navy when he took the post, he was uncomfortable with Washington life, with the idea that business extended into the social realm and that officials were expected to be "on" all the time. Nevertheless, *Business Week* could call Matthews "the most underestimated man in Washington" in its September 9, 1950, issue.

Matthews's first year or so as Secretary of the Navy was uneventful. His challenge had been to put down the "revolt of the admirals." Career officers unhappy with Truman's plan for unification of the armed forces felt the plan diminished the influence of the Navy. Matthews fired the leading dissident, Admiral Louis E. Denfeld.

Other than that, his most controversial quality was a penchant for permitting Navy personnel and property to show up at private festivities at public expense. A squadron of Navy destroyers was sent to a Knights of Columbus convention in Portland, Oregon. Naval Air Reserve planes flew hundreds of miles just to scatter roses over a procession in honor of St. Theresa in Pennsylvania.

Matthews never wavered on his hatred of communism. The purple language of the following, from a speech Matthews wrote and delivered at a Jesuit Mission benefit dinner in New York (November 9, 1950), is almost chilling. Communism, he said:

> *. . . is the fatal mischief of man's fallen nature exerting its nefarious influence in a new and falsely alluring garb. Beneath the sophistry of its superficial philosophy, however, are visible the familiar devices to deceive men's minds and corrupt their wills that have been the Devil's handmaidens throughout his foul career. No slightly effective resistance can stop the encroachment of this pernicious influence on the thinking and behavior of the peoples exposed to its withering effects. Two phases of opposition are called for: Its advance must be blocked; its misguided apostles must be converted.*

AFTERMATH

Matthews's Boston speech set in motion a complex series of outcries, commendations, clarifications, and denials. Immediately after Matthews's speech, Labor Secretary Tobin assured reporters that Matthews "speaks as the official representative of President Truman." This was untrue. The Administration was mortified when it learned of Matthews's comments.

Secretary of State Acheson conferred with Truman and issued a curt statement to the press: "Secretary Matthews' speech was not cleared with the Department of State, and his views do not represent United States policy. The United States does not favor instituting a war of any kind." Truman chided Matthews by phone. Ambassador Philip Jessup, one of Acheson's advisers, told the press that "dropping atomic bombs on Moscow is not the way America does things."

The State Department had never been on friendly terms with Matthews, with his claims of Communists in their midst. The press quoted an unnamed State Department official as saying the speech "played right into the hands of the Russians." "Its propaganda effect abroad can be very bad," the State Department source added. Congressman John Kee of West Virginia, chairman of the House Foreign Affairs Committee, sniped that Administration officials whose expertise did not lie in the realm of foreign policy might best "keep their big mouths shut."

A *New York Times* editorial (August 27, 1950) refused to believe Matthews meant what he said: ". . . we hardly think he could have intended to urge an aggressive war on the part of the United States." The *Chicago Tribune* Press Service cited "sources close to Mr. Matthews" as saying that the Navy Secretary "still believes the United States should hit first in a war with Soviet Russia, but he will keep mum until President Truman gives him the word." Its sources claimed that Matthews "did not feel that his speech contradicted State Department or White House policy."

On August 28, Matthews spoke to a reporter and tried his best to deny that he was saying much of anything. He said that "I wasn't intending to be speaking for anybody but myself. The speech speaks for itself. I don't say that we should institute a war to compel cooperation. I said that we might have to. That's all. I don't advocate that now, of course."

Matthews's office log shows that he met with the President for a half hour on September 18. By this time the Washington grapevine was whispering that Matthews would be asked to resign. Dan Kimball was rumored to be in line for Matthews's job. Matthews said of the meeting that Truman reiterated that he didn't want him to resign, and that this was the second reassurance in ten days.

Meanwhile, people who had been thinking along the same lines as Matthews began to speak out. The first was the commander of the American Legion, George N. Craig. The American Legion has little to do with American foreign policy, but Craig was speaking at a meeting in Washington the night after Matthews's speech. The press was eager for a follow-up story on preventive war, and Craig's remarks got more media attention than they would have under almost any other circumstances. Craig suggested that the United States extend the Monroe Doctrine to the entire world. The United States should outlaw communism and institute universal military training. "If Russia is going to bring on World War III, let us have it on our own terms. If Russian puppets start trouble anywhere . . . that will be the signal for our bombers to wing toward Moscow." Craig, unlike Matthews, specifically mentioned atomic bombs and the Soviet Union. "America must now take a resolute stand for world peace by compulsion. We have this preventive power. We have the atomic bomb and the industrial might. We can and must put our manpower behind both."

A handful of people in positions of power spoke up in favor of preventive war. Senator Richard B. Russell of Georgia told Matthews his speech was "a perfectly grand thing." Russell allowed, "It was high time for somebody in high position to make a statement of that kind and wake up the American people and get them thinking in broader terms."

Russell was joined in praise by two other senators, Karl E. Mundt (South Dakota—a Republican) and Elmer Thomas (Oklahoma). All were politic enough to describe preventive war as "something the American people should be thinking about" (Russell's words); something we might have to do one day—not right this instant.

The Soviets promptly accused the United States of warmongering, an unexceptional charge, all things considered. On August 29, Bucharest radio broadcast a choice example of cold-war rhetoric:

The criminal plans of the American imperialists have again been revealed through the imprudence of U.S. Secretary of the Navy Matthews, who stated that the United States must declare war to impose peace. This cynical statement has provoked the indignation of U.S. public opinion. For this reason the State Department hurried to disapprove of Matthews' statements because the boner expressed the thoughts of the Wall Street cannibals.

Soviet Air Force General Vassily Stalin, son of Joseph, responded with the unconvincing assertion that no enemy bomber could ever reach any Soviet target, no matter how fast or how high it flew.

PUBLIC REACTION

The Boston speech also opened a Pandora's box of American public opinion on the nuclear dilemma. Both Matthews's and Truman's offices saved the letters they received on the preventive war speech, and this correspondence is now preserved in the Truman Library, Independence, Missouri. Matthews's staff made an analysis of the letters he received and reported 107 letters in favor of the Boston speech vs. 55 against. However, the letters addressed to Truman were far more negative than positive; more writers overall were against preventive war than for it.

Many of the writers pro and con were utter cranks; the letters do not make a strong case for the rationality of the American people. An eighty-six-year-old woman with the Socialist party of Maine composed an entire letter in verse ("'AGGRESSION' for peace you talk too loud/ 'Aggression' should not be allowed/To put so much LIFE in a shroud and MOTHERS in grief and mourning bowed . . .").

Those for Matthews were strongly for him. An Ozone Park, New York, man found Matthews's remarks "the most profound I have ever read in my lifetime." A Matthews supporter in Gardena, California, wrote:

> *I served in Naval Operation W.W.I. My older boy was in Japan W.W.II. Now my younger boy is of draft age. Just how long is this to go on? Now that we have the H Bomb advantage shoot the works! . . . Give em Hell. They are asking for it!*

A Long Beach, California, writer said,

> *We are 110 percent back of you in your idea for bombing Stalin. When on the farm we wanted to get rid of the skunks killing our chickens we saught [sic] out their den and blasted that; so with that Skunk Stalin, blast him with the bomb and give him what is coming to him.*

This letter, signed "Yours for a quick victory," was from a Presbyterian minister.

A lot of the letters demanded Matthews's resignation or asked Truman to fire him. "Resign—you maniac" is the complete text of a postcard addressed to Matthews from a Seattle man. The American Socialist party asked for his resignation, as did the "Housewives of Laurelton, NY." "Please resign and submit to a psychiatric examination," a San Leandro, California, housewife counseled. A Minneapolis man wrote, "Did it not occur to you that this irresponsible pronouncement of yours might provide the Commies with a provocation to attack us. Again—for the good of the nation please resign & pronto!" A Philadelphia woman scolded Truman, "You had better fire him or I am going to think you and the rest of the Democrats secretly think what he does . . ."

A Durham, North Carolina, man objected to Matthews's phrase, "compel cooperation," pointing out that "Cooperation is the one relationship you cannot compel." A Hartsdale, New York, man said that "as a veteran of the Second World War, I know that all we may be sure that any war will prevent is peace."

"A preventive war is certainly one of Satan's own plans; it is strange and fearful to hear it mouthed by one who is supposed to be an ardent churchman," said a Los Angeles woman. Numerous writers referred Matthews to specific Bible passages—usually with the intent of convincing him that preventive war was morally wrong, though a few saw nuclear holocaust as a necessary fulfillment of biblical prophecy. A Schroeder, Minnesota, writer recommended that Matthews read Tolstoy's *War and Peace,* from which he would learn that "The Russians will fight like demons for Mother Russia."

A distressingly large proportion of the letters against the speech were from anti-Catholic bigots who were at least as interested in hinting at dark Vatican conspiracies as in opposing preventive war. A Lemon Grove, California, woman wrote that the speech "points up for true Americans with no foreign allegiance the danger of having Catholics in public office." A Pasadena, California, woman asked, "Was Secretary Matthews, who I understand is a zealous Catholic, expressing the desire of the *Vatican* to have us take on all out war with Russia?"

Matthews was likened to Hitler in a number of letters; Stalin, Mussolini, and the Spanish Inquisition were also mentioned. He was called "the curse and scourge of the world" (Santa Fe, New Mexico); "You sir are a fool *below* fools," another wrote (Tucson, Arizona); "What meat for Malik in the Security Council!" a New York man commented in reference to the Soviet UN representative known for his claims of U.S. warmongering. A Mount Holly, New Jersey, man told Matthews: "If war should come you could best serve the country by piloting a guided missile to its target personally."

People on both sides of the issue claimed that nearly everyone shared their opinion. The Monday after the speech, a caller assured Matthews that his speech expressed the views of "about 90 people out of every 100 throughout the length and breadth of this land." A New York man insisted that "I personally know no one who would be willing to 'institute a war to compel cooperation for peace,'" while a writer across town in Jamaica, New York, reported that "I and all my associates subscribe to your views. The next time any Russian satellite starts trouble, send our bombers to Moscow."

William Loeb, influential publisher of the *Manchester* (New Hampshire) *Evening Leader,* wrote Matthews to tell him that "up here we are all for you. Someone must tell the ugly truth, or as a nation we will be in the position of the man who feels so very comfortable in the snow bank when it is 40° below zero that he never wakes up until it is another world."

A minister in Hyde Park, Massachusetts, said, "I think your suggestion for a war of aggression in the interest of peace is un-Christian and brutal; and I think I speak for many leaders and people in our churches." From Little Rock, Arkansas, came assurance that ". . . your viewpoint is shared by a great many people, including myself—wonder why Truman don't just give Stalin the 'keys to the city' and be done with it."

It is hard not to be amazed at the number of sincere people who wrote Truman insisting on preventive war's near-unanimous support with the American public, and implying that only a few inside-the-beltway politicians could possibly disagree. A Wrightsville, Pennsylvania, doctor wrote Truman, "The speech of Mr. Matthews, Navy Secretary, was like a fresh breeze in the close air of the short-sightedness of American statesmen. He told exactly what most of us either thought or felt." A Watertown, New York, woman told the President, "I hear many people talking—all say 'put the bomb on the Kremlin' and let's get it over. As I see it—to use an old saying, 'It's either fish or cut bait.' " From Tucson, Arizona: "From what I can gather as to the opinion in this area, his preventive war plan is very popular and merely an expression of a very widely held belief. If any cabinet member stands in the way, he should be fired." And a Buffalo, New York, man wrote on his corporate letterhead:

> *A great many of us are tired of playing peek-a-boo with facts, tired of timidly hinting at reality in diplomatic phrases. Among my company's workers and those in other plants (in which we work) I have yet to find a man who does not want a showdown with Russia. In the little shop groups at rest periods or lunch hour you hear someone say, "We ought to lick Russia as soon as we can get ready for it. To hell with wondering where they will start something next." The others nod their heads.*

Finally, some felt Matthews didn't go far enough. They suggested bombing China; rounding up American Communists and placing them in road gangs; electrocuting Communists.

WAS IT A TRIAL BALLOON?

Sensational comments by high-ranking officials are tough to repudiate. One of the unanswered questions of cold-war history is whether Matthews's speech was a trial balloon. "Could this be the new, post-Korea diplomacy?" *Newsweek* magazine wondered. The *Washington Star* wrote (August 27, 1950), "Whether there is another group in the Cabinet for which Mr. Matthews has now become spokesman, whether his argument reflected Navy or general military thinking—those were questions which had diplomats guessing." The same day's *Washington Times-Herald* asserted that "There are many persons in Congress and in the govern[ment] who favor dropping atom bombs on the Soviet Union . . ."—so many, apparently, that it was unnecessary to name any. *Washington Post* columnist Marquis Childs wrote (August 31): "No one who has even a slight knowledge of what is actually happening here can believe that he [Matthews] thought up the idea of preventive war. But now he has been made to confess the blame and stand in the corner."

The press reported rumors that Matthews's boss, Secretary of Defense Louis Johnson, had discussed preventive war in private talks, and that Harold Stassen had seriously contemplated the issue.

Matthews *was* voicing ideas that had been kicked around for some time. That's not the same thing as being the mouthpiece of an organized preventive-war faction. No convincing evidence has turned up that Matthews spoke for an organized group, and the notion that he was expressing Truman's ideas must be ruled out.

THE MACARTHUR SPEECH

Another embarrassing flap erupted on August 27. General Douglas MacArthur had prepared a statement to be read at the Veterans of Foreign Wars convention in Chicago. MacArthur's statement claimed that if Formosa fell into the hands of an enemy of the United States, "any future battle area" would be shifted "5,000 miles to the eastward to the coasts of the American continents, our own home coasts."

The MacArthur statement argued for a hard line in Formosa, citing archaic views of the "Oriental mind." Said MacArthur, "Nothing could be more fallacious than the threadbare argument by those who advocate appeasement and defeatism in the Pacific that if we defend Formosa we alienate continental Asia. Those who speak thus do not understand the Orient. They do not grasp that it is in the pattern of Oriental psychology to respect and follow aggressive, resolute and dynamic leadership—to quickly turn from leadership characterized by timidity or vacillation—and to underestimate the Oriental mentality."

Shortly before the statement was to be made public, MacArthur sent a telegram to VFW National Commander Clyde A. Lewis: "I regret to inform you that I have been directed to withdraw my message." Truman had told MacArthur to do so.

MacArthur had already sent the message to a number of press sources. It had been printed in *U.S. News and World Report,* and copies of that issue had been put in the mail to subscribers. Thus the withdrawn statement was quickly picked up and reprinted in newspapers. Republicans read it into the *Congressional Record.*

ORVIL ANDERSON

Still another controversy was brewing. This time the talk came from a respected but low-ranking major general in the Air Force. At the age of fifty-five, Major General Orvil A. Anderson had compiled an impressive service record. He had flown in both world wars and had been a deputy commander of the Eighth Air Force in the second. Between wars he took part in a 1935 experiment sponsored by the Army Air Corps and the National Geographic Society. Wearing little more than fighter pilot suits and football helmets, Anderson and a partner steered the *Explorer II* balloon to a world-record height of 72,395 feet (almost 14 miles). This remarkable ascent into the stratosphere made him a hero comparable to the astronauts of later generations. The record still stood in 1950. It was not shattered until 1951, and then by a plane.

That said, Anderson was not the sort of major general that the Pentagon consulted on matters of policy. He was comfortably installed as a lecturer at the Air Force's Air War College at Maxwell Field, Alabama. A reporter for the *Montgomery Advertiser* had heard that Anderson's lectures included speculations about an atomic attack on the Soviet Union. Sensing a good story in the aftermath of the Matthews affair, the reporter asked Anderson for an interview. Anderson agreed —and turned out to be a live one.

"To assume that Russia won't use their A-bombs if we sit by and watch them build them is a dangerous assumption," Anderson said. His words border on the grotesque:

> *We're at war, damn it. I don't advocate preventive war, I advocate the shedding of illusions. I advocate saying to Stalin: "Joe, you're not kidding anybody. You say you are going to destroy us." And if he says, "yes"—and he has been saying "yes" all the time—we must conclude civilization demands that we act.*
>
> *Give me the order to do it and I can break up Russia's five A-bomb nests in a week. And when I went up to Christ, I think I could explain why I wanted to do it—now—before it's too late. I think I could explain to him that I saved civilization.*

Five A-bomb nests? It was not common knowledge that the Soviets had five bomb sites. Had Anderson spilled classified information? The question was put to him later, and Anderson said he had just picked the number out of the air. That's not inconceivable, but it is also the sort of thing that a person who had let a military secret slip would *have* to say.

At about the same time (August 30), still another general, Albert C. Wedemeyer, discussed preventive war in a talk at the National War College in Washington, D.C. The public little knows one general from another. All of a sudden, it seemed as if the American military establishment was abuzz with talk of preventive war.

Journalists soon learned that Maxwell Air Force Base had been a hotbed of preventive war talk for some time. During the Berlin blockade, Maxwell's former commander, Brigadier General S. D. Grubbs, had told a Montgomery civic group that the United States ought to demand the Soviets lift the blockade in thirty-six hours or face atomic attack.

It was also reported that Anderson had spoken in favor of an immediate war with the Soviet Union at a Kiwanis Club meeting some time previously, with no untoward reaction. It was beginning to look like bombing Moscow was almost a cliché on the service club circuit. In the *Washington Post,* on August 31, columnist Drew Pearson cited "concrete evidence that the general follows a deliberate program at the Air

College aimed to indoctrinate students with the idea of an immediate attack."

The Air Force acted quickly. On September 1, Air Force Chief of Staff Hoyt Vandenberg suspended Anderson from his position as commandant of the Air War College pending a formal investigation. He stated, "The Air Force, first, last and always, is primarily an instrument for peace."

The same day, Truman did his best to set the record straight. He gave a radio and TV fireside chat on Korea and preventive war. Truman wore a blue necktie patterned with the United Nations flag and tiny oak-wreathed globes. "My fellow Americans," he began, "tonight I want to talk to you about Korea, about why we are there and what our objectives are."

The President held the Soviet Union responsible for the Korean conflict and for a general policy of imperialism. He advocated doubling the size of the armed forces (from 1.5 million to 3 million) to meet this threat. He emphasized that the United States had no designs on Korea or Taiwan.

Truman said, "We do not believe in aggressive or preventive war . . . We are arming only for defense against aggression." He called preventive war "the weapon of dictators, not of free democratic countries like the United States."

PRESS REACTION

Despite Truman's disavowal, stories on preventive war filled the press for weeks afterward. "Talk of using the A-bomb is heard as it has never been heard before," reported *Life* (December 11, 1950). "When a man knows he has a good chance to be A-bombed, nothing can stop him from wondering whether there isn't something he can do to prevent it," observed *Time* magazine's September 18, 1950 issue. ". . . Very few Americans now believe that the Kremlin can be conciliated or appeased or reasoned with. Very few are content to sit back and wait for the Communists to strike."

Most publications came out strongly against preventive war. *Time*'s editors concluded that "Militarily, preventive war by the U.S. in 1950 would be a blunder of tragic proportions. . . . In such a situation the question of the morality of preventive war, which troubles many

Americans, may not even arise. Whether or not preventive war is morally bad, the facts of 1950 make it military nonsense."

"That there has been discussion of preventive war within the military establishment goes without saying, just as the subject has been argued by countless thousands of civilians," ran a guest editorial by retired U.S. Air Force General Carl Spaatz in *Newsweek* (September 11, 1950). "It has been frequently said of late that the theory of preventive war is to 'do unto others what you fear they will do unto you—but do it first.' This is the thinking of the weak and fearful. It is gangster reasoning, and we are certainly not a trigger-happy nation."

Collier's magazine (November 11, 1950) wrote:

> *For several months we have been reading and hearing about something called "preventive war." We have examined the arguments in its favor. We have listened to them and thought about them. And in the end we haven't even learned what the term means. "Preventive war." What would it prevent?*
>
> *It wouldn't prevent war.*
>
> *It wouldn't prevent the atomic bombing of the United States. It takes an incredible optimism to believe that the Russians have collected all their atomic weapons in one vulnerable place, and that the U.S. Air Force knows exactly where that place is, and that our bombers would surely and inevitably get through on their first strike and destroy all the Soviet bombs on the first day of war. . . .*

The *Pilot,* newspaper of the Catholic Archdiocese of Boston, asked whether preventive war was morally correct. Their answer was "yes" —provided it was fought for a "morally certain right" and that other solutions had failed. The *Pilot* said, "There is considerable evidence to establish the fact that the Soviet is guilty of real crimes and is contemplating further ones, and we seek merely to defend basic human rights." The *Pilot* concluded that a preventive war against the Soviet Union might be necessary.

Preventive war was the subject of at least one book, *Half Slave, Half Free* (1950) by Hallett Abend. Abend theorized that the Soviets were even then smuggling bombs in pieces into U.S. cities and assembling them. "What could our government do if some bright day the Soviet ambassador submitted terms for abject surrender, together with a threat that if capitulation were not signed within the hour, the secreted bombs would be exploded simultaneously with silent timing devices?" The author hinted a nuclear ultimatum was the plan of unnamed "high military officers" and civilian leaders. Abend discusses a surprise atomic attack, judging it a "dastardly move . . . even though it might promise an early and cheap victory."

HOW MANY BOMBS?

In a Western movie, the gunslinger who draws late dies instantly. Reality is bloodier; the dying gunslinger may get off a few good shots at his speedier opponent. Oddly enough, one of the least-mentioned issues in the preventive war debate was whether it would work. One reason for this was the profound ignorance, even among decision-makers, about the size of the nuclear stockpile.

The number of atomic bombs on hand was a secret guarded almost to point of absurdity. According to Truman (who was not formally briefed on the matter until April 1947, when he had been President two years), there was no document anywhere in Washington, that city of paperwork, saying in plain English how many bombs the United States had. The numbers were said to be "recorded on detached pieces of paper safeguarded in a special way."

A famous story claims that Truman was shocked to learn how few bombs there were. This refers to the formal briefing on April 3, 1947. Various sources claim Truman learned there were components for only seven bombs. If this came as a surprise, it must have been because progress had been so slow. Six months earlier, in October 1946, Truman told staffers that he did not believe there were more than a "half dozen" bombs available. But Truman believed "that was enough to win a war."

Hardly anyone was told the secret, even among those making military policy. In January 1949, Senator Brien McMahon, reportedly a proponent of preventive war, complained that Congress was like

> *a general who must train his troops without knowing how many rounds of ammunition they will be issued. When we debate the necessity of a 65,000-ton aircraft carrier, or a 70-group air force or universal military training, I fear that we quite literally do not know what we are talking about. We do not know how many*

atomic weapons we possess, and therefore, I fear we lack perspective to pass upon any major defense issue.

It was evidently a matter of pride with General Leslie Groves that the number of people knowing this secret actually grew less as the years passed. In 1947 Secretary of the Navy James Forrestal and Chief of Naval Operations Admiral Chester Nimitz were asked to recommend production rates for atomic weapons. Each assumed the other knew how many bombs there were; neither did. Groves stonewalled even when asked by General Curtis LeMay in 1947 (a year before LeMay became commander of the SAC). Groves replied that "that information is quite complicated and is based on many factors. I cannot answer your question because I force myself to forget the numbers involved."

Actually, Groves was not being so evasive as his words may sound. There is no simple answer to the question of how many bombs America had, even were all the facts known.

A completely assembled "Fat Man" bomb was a perishable commodity. It remained functional for only about forty-eight hours before its batteries ran down. Then it had to be partially disassembled so the batteries could be recharged. Each bomb further required an "initiator" of polonium 210, an extremely unstable isotope with a half life of only 138 days. These had to be replaced every few months. (The principle nuclear fuel, plutonium, is more stable.) Thus the components of bombs were stored separately, to be assembled if and when needed. Great expense and effort went into maintaining the stockpile of components—particularly the polonium initiators.

The question, then, is not how many complete bombs existed, but how many could have been assembled if needed. The bottleneck is whatever crucial component is in shortest supply.

An article by historian David Alan Rosenberg in the May 1982 *Bulletin of the Atomic Scientists* revealed just how small the atomic arsenal was. According to the Department of Energy, in mid-1945, the United States had nuclear cores and mechanical assemblies for just 2 atomic bombs.[1] The cores and assemblies for 9 bombs existed in 1946,

1. There is some confusion about this figure. It appears to refer to the one exploded at the Trinity Test in New Mexico and the one dropped on Nagasaki. It does not count the Hiroshima bomb, which apparently wasn't done at the time. If that is so, the United States may have had no atomic weapons at all immediately after Nagasaki. The Depart-

13 in 1947, and 50 in 1948. The Department of Energy did not give numbers of polonium initiators, which were probably the bottleneck in the early years. The figure of 7 bombs at the time of Truman's April 1947 briefing presumably means that the production of polonium was enough to arm only 7 of the 13 or so nuclear cores in existence at that time.

The Department of Energy still classifies the number of bombs after 1948. It is evident that the numbers rose dramatically, though. The Department of Energy told Rosenberg that the number of mechanical assemblies for atomic bombs was 240 in 1949 (compared with 55 in 1948). In 1950, it was 688. The department said that there were more assemblies than nuclear cores in these years. (By comparison, the 1990 START treaty ratified by Presidents Bush and Gorbachev provided for both sides to *reduce* their stockpiles to 6,000 nuclear warheads each.)

There are other clues to the size of the arsenal. Rosenberg cites a 1950 study on strategic air operations that claims a hypothetical air offensive involving 292 atomic bombs was possible that year. That figure sets a lower limit on the bombs available in 1950. The growth of the atomic arsenal apparently looks something like this:

	Complete Bombs Available (as of mid-year)
1945	about 2
1946	about 6
1947	about 7
1948	no more than 50
1949	less than 240
1950	between 292 and 688

It's even harder to get reliable information on the Soviet arsenal. A story in the July 8, 1950, issue of *Business Week* claimed the Soviets had about 10 atomic bombs. A few months later, *Time* magazine put the number at "more than 10 and less than 60—enough to give the Kremlin a means of dreadful retaliation."

These figures are only part of the story. There were relatively few of the modified planes capable of carrying an atomic bomb. Bomb assembly teams and flight crews were also in short supply. The 509th Bomb

ment of Energy's historian told Rosenberg that the figure for 1945 may have referred to the end of the calendar year (December 31).

Group at Walker Air Force Base, Roswell, New Mexico, was the only unit capable of delivering a bomb until mid-1948.

It would be difficult to deliver the bombs on target, and some U.S. tests on this matter were not encouraging. In a 1947 SAC drill, almost half the bombers failed to get airborne. In 1948, SAC staged a nighttime mock attack on Dayton, Ohio, and not a single bomber managed to complete its mission as ordered.

Added to this was the poor defense mapping of the Soviet Union. The official Soviet maps were purposely distorted to confuse would-be attackers. The Strategic Air Command resorted to aerial photographs of Russia captured from the Nazis and even maps dating from Czarist times. The upshot is that, in a hypothetical atomic attack on the Soviet Union circa 1950, many bombs would not hit their targets.

A single fission bomb could lay waste to the Kremlin (which occupies a little more than a tenth of a square mile) and much of historic central Moscow as well. Nevertheless, the city of Moscow covers 386 square miles. Accepting the figure of 4 square miles as the zone of more or less complete devastation, most of the population, buildings, and industry of Moscow would be relatively unscathed by a single bomb.

For 1950, the year in which preventive war came to the forefront, there were anywhere from 292 to 688 bombs. It might have been realistic, then, for the United States to contemplate an attack in which as many as several hundred targets would be destroyed. A thousand square miles or more might be laid to waste, an area approaching that of Rhode Island.

It is far more difficult to predict the consequences of an attack. The press sometimes referred to preventive war in such terms as "bombing Russia back to the stone age." There was a tendency to think that an atomic (fission bomb) attack would simply annihilate the Soviet Union. This clearly wouldn't have been the case. As a 1947 article in *United States News* put it, "Russia is spread out, has no one vital nerve center. Use of enough bombs might kill a lot of people, might knock out Moscow and some steel mills, but would not win a war on the basis of anything now known. Such attacks are more likely to unite the Russian people."

The apocalyptic specter of a great nation wiped off the map is all too possible in this age of arsenals of thousands of hydrogen bombs. But it was never a possibility in the preventive war era. Even in 1950, America didn't have enough bombs.

CODA

The results of the Anderson investigation were never made public. Anderson was reassigned to command of the 3750's Technical Training Wing at Sheppard Air Force Base, Wichita Falls, Texas. An Air Force spokesman insisted that this was not a demotion, that Anderson had not been disciplined or reprimanded in any way for his remarks. Anderson evidently didn't share this happy view of the matter. He applied for retirement rather than accept the new post.

The preventive war speech came to haunt Matthews. He resigned as Secretary of the Navy the following year. There's little indication that the Administration was sad to see him go, and he *was* succeeded by Dan Kimball after all. Matthews accepted the ambassadorship to Ireland. While on vacation at his Omaha home, Matthews died unexpectedly of a heart attack on October 18, 1952. The Rotary Club observed a minute of silence in his memory. The in-house publication of one of Matthews's businesses eulogized, "He Embodied All that is Good in Mankind."

Talk of preventive war persisted through the early 1950s. Some of it was spurred by the successful development of the H-bomb, which briefly seemed to restore a U.S. advantage. A 1952 rumor held that Air Force Secretary Finletter had said of the H-bomb: "With seven of these weapons, we can rule the world." The statement was supposed to have been made at a secret Pentagon briefing. Finletter and others denied the comment. How to prevent World War III—prevent it without starting it—became a popular grade school topic in 1951 and 1952.

Like Truman, Eisenhower rejected preventive war. But well into the Eisenhower administration, Winston Churchill was still talking about preventive war. "Churchill, debating with himself over a whisky and soda in a lakeside garden about which of his enemies he proposed to annihilate if the atomic bomb came into his possession, would be a subject for high comedy if it were not so dreadfully and desperately tragic," declares Robert Payne's 1974 biography, *The Great Man.* Britain tested its first bomb in the Australian desert in October 1952. Payne's book records a strange dream Churchill had at about this time. Churchill found himself in a train, travelling across Russia with Molotov and Voroshilov. This unlikely trio was staging a counterrevolution. Churchill had atomic bombs the size of matchboxes. All of Russia was laid to waste, the Russian people exterminated.

In December 1953, Churchill met with Eisenhower and Premier Laniel of France in Bermuda. Churchill's personal physician, Lord Moran, traveled with the Prime Minister and kept a diary. The entry for December 3 records a conversation between Churchill (P.M.) and an unnamed professor:

> *P.M.: There was a time when the Western powers could have used the bomb without any reply by Russia. That time has gone. How many atomic bombs do you think the Russians have?*
> *PROF: Oh, between three and four hundred. The Americans may have three or four thousand.*
> *P.M.: If there were war, Europe would be battered and subjugated; Britain shattered, but I hope not subjugated. Russia would be left without a central government and incapable of carrying on a modern war.*

Moran's entry for December 5 not only has Churchill discussing preventive war (as a devil's advocate?) but claims, at second hand, that Eisenhower did not rule it out. The entry reads in part:

> *The P.M. is less sure about things today. It appears that when he pleaded with Ike that Russia was changed, Ike spoke of her as a whore, who might have changed her dress, but who should be chased from the streets. Russia, according to Ike, was out to destroy the civilized world.*
> *"Of course," said the P.M., pacing up and down the room, "anyone could say the Russians are evil minded and mean to destroy the free countries. Well, if we really feel like that, perhaps we ought to take action before they get as many atomic bombs as America has. I made that point to Ike, who said, perhaps logically, that it ought to be considered. But if one did not believe that such a large fraction of the world was evil, it could do no harm to try and be friendly as long as we did not relax our defense preparations."*

In 1957 Eisenhower sent Robert Sprague and Jerome Weisner to Omaha to confer with SAC head Curtis LeMay. They told the general that few SAC bombers could survive a surprise Soviet attack. As

Sprague and Weisner recounted the meeting *(Los Angeles Times,* July 24, 1989), LeMay agreed with their conclusions. He was counting on U-2 spy planes flying over the Soviet Union to give him a week's warning of preparations for a surprise attack. "I'll knock the shit out of them before they get off the ground," LeMay boasted. When Sprague and Weisner objected that first use of nuclear weapons was not national policy, LeMay said, "No, it's not national policy. But it's my policy."

8

GAME THEORY AND ITS DISCONTENTS

Slowly the original group of game theorists at RAND broke up. Von Neumann became busier and had less time for game theory. In 1951 RAND doubled von Neumann's daily pay to $100 in an effort to get him to spend more time with them. It didn't have much effect. Von Neumann finally dropped his association with RAND in early 1955, when his appointment to the Atomic Energy Commission forced him to curtail outside work.

Melvin Dresher was one of the few who remained at RAND until his retirement in the 1980s. Merrill Flood left in 1953 for Columbia University, where he helped recruit RAND colleague R. Duncan Luce. For Flood, this was the start of a diverse career in academia and elsewhere. As a consultant to a television producer, Flood helped revise the cash payouts on the quiz show "Joker's Wild." In recent years, his interests have included the mathematics of voting. He has explored, and attempted to popularize, voting systems that more fairly represent minority interests.

John Nash became increasingly paranoid. He would pester his colleagues with peculiar ideas for tightening security at RAND. He was eventually committed to a psychiatric hospital for treatment. He recovered and joined the Institute for Advanced Study.

CRITICISM OF GAME THEORY

Views on game theory were changing. A decade after the publication of *Theory of Games and Economic Behavior,* there was a correction to the early euphoria. Game theory was deprecated, distrusted, even reviled.

To many, game theory, ever intertwined with the figure of John von Neumann, appeared to encapsulate a callous cynicism about the fate of the human race. A few examples will show the severity of this reappraisal. In a 1952 letter to Norbert Wiener, anthropologist Gregory Bateson wrote:

> *What applications of the theory of games do, is to reinforce the players' acceptance of the rules and competitive premises, and therefore make it more and more difficult for the players to conceive that there might be other ways of meeting and dealing with each other. . . . its use propagates changes, and I suspect that the long term changes so propagated are in a paranoidal direction and odious. I am thinking not only of the propagation of the premises of distrust which are built into the von Neumann model* ex hypothesi, *but also of the more abstract premise that human nature is unchangeable. . . . Von Neumann's "players" differ profoundly from people and mammals in that those robots totally lack humor and are totally unable to "play" (in the sense in which the word is applied to kittens and puppies).*

Game theorists were well aware of their profession's tarnished image. In 1954, RAND's John Williams wrote that game theorists "are often viewed by the professional students of man as precocious children who, not appreciating the true complexity of man and his works, wander in wide-eyed innocence, expecting that their toy weapons will slay live dragons just as well as they did inanimate ones."

Or try RAND alumni R. Duncan Luce and Howard Raiffa in their 1957 book *Games and Decisions:* "We have the historical fact that many social scientists have become disillusioned with game theory. Initially there was a naive band-wagon feeling that game theory solved innumerable problems of sociology and economics, or that, at the least it made their solution a practical matter of a few years' work. This has not turned out to be the case."

Charles Hitch, head of RAND's economics division, told *Harper's* magazine in 1960: "For our purposes, Game Theory has been quite disappointing."

The public was apt to think of game theory—if it thought of it at all —as a tool for justifying nuclear war. Nuel Pharr Davis's *Lawrence & Oppenheimer* quotes J. Robert Oppenheimer as asking (1960), "What are we to make of a civilization which has always regarded ethics an essential part of human life, and . . . which has not been able to talk about the prospect of killing almost everybody, except in prudential and game-theoretic terms?"

Others recognized that much of the problem rested with the mind-set of those using game theory. In a 1962 article on "The Use and Misuse of Game Theory" for *Scientific American,* Anatol Rapoport wrote perceptively:

> *. . . game theory has been embraced in certain quarters where Francis Bacon's dictum "Knowledge is power" is interpreted in its primitive, brutal sense. The decision makers in our society are overwhelmingly preoccupied with power conflict, be it in business, in politics or in the military. Game theory is a "science of conflict." What could this new science be but a reservoir of power for those who get there fastest with the mostest? A thorough understanding of game theory should dim these greedy hopes.*

Misgivings continued well into the 1980s and perhaps to the present day. Steve J. Heims, in his *John Von Neumann and Norbert Wiener* (1980) wrote: "Game theory portrays a world of people relentlessly and ruthlessly but with intelligence and calculation pursuing what each perceives to be his own interest. . . . The harshness of this Hobbesian picture of human behavior is repugnant to many, but von Neumann would much rather err on the side of mistrust and suspicion than be caught in wishful thinking about the nature of people and society."

The charges against game theory fall into two broad categories: that game theory is a Machiavellian exercise to justify war or immoral acts; and that game theory isn't useful in the real world. (The purely mathematical validity of game theory has never been in question.) Both objections are worth examining.

UTILITY AND MACHIAVELLI

The players of game theory are presented as "self-serving egoists." Descriptions of the prisoner's dilemma, including Tucker's and others in this book, ask you to put yourself in the place of an amoral, hardened outlaw dealing with an equally ruthless adversary. Why such hard-boiled stories?

It's *not* because game theory is about how people with a certain psychology (self-centered and ruthless) play games. It's a matter of descriptive economy. Game theory talks only about utility—not about years in prison or dollars or anything familiar like that. You will recall that utility is an abstraction that can be thought of as the "points" that players play for. Since utility is an unfamiliar concept, science writers try to explain game theory in such a way that they can omit a digression explaining utility. This is possible when there is a simple and obvious correspondence between utility and dollars, years, or some other physical unit.

For an amoral egoist, the correspondence *is* simple. Money is good—the more of it, the better! It's much easier to talk about people whose utility corresponds to a tangible object, with no ethical scruples about betrayal or psychic rewards for benevolence to complicate the picture. The mistake is in thinking that game theory is specifically about such people.

Like arithmetic, game theory is an abstraction that applies to the real world only to the extent that its rigorous requirements are met. Consider the person who is calculating the value of the change in her pocket. She counts out three pennies and seven nickels, and computes that she has thirty-eight cents. Then she finds out that she has miscounted and there are only six nickels. Does this mean her arithmetic is wrong? Of course not. If you count wrong, you can't blame arithmetic. Likewise, assessment of utilities is a prerequisite for applications of game theory.

Here arithmetic and game theory diverge. Any two people who accurately count a group of pennies must come to the same result. Utility, though, is subjective by definition. Any two people are likely to rank a set of game outcomes differently, provided the outcomes are not cash prizes but very complex states of human affairs.

Game theory is a kaleidoscope that can only reflect the value systems of those who apply it. If game theoretic prescriptions sometimes seem Machiavellian, it is generally because the value systems of those who applied the game theory are Machiavellian.

There is also a crucial difference between zero-sum games with a saddle point and those without. A saddle point exists even if the outcomes are merely ranked in order of preference. For other games, utilities must be placed on a strict numerical scale (an "interval scale"). Otherwise, there are no numbers from which to compute the probabilities for the correct mixed strategy.

It is difficult to have much confidence that military strategists can, as a practical matter, assign numbers to outcomes like peace, limited war, and nuclear holocaust. One can pull "reasonable" numbers out of the air, to be sure, but that defeats the purpose of applying game theory, which was to provide recommendations more accurate than intuition. As Rapoport noted in *Scientific American:*

> *Unless this more precise quantification of preferences can be made, rational decisions cannot be made in the context of a game without a saddle point. I have often wondered to what extent decision makers who have been "sold" on game theory have understood this last verdict of impossibility, which is no less categorical than the verdict on squaring the circle with classical tools. I have seen many research proposals and listened to long discussions of how hot and cold wars can be "gamed." Allowing for the moment that hot and cold wars are zero-sum games (which they are not!), the assignment of "utilities" to outcomes must be made on an interval scale. There is the problem. Of course this problem can be bypassed, and the utilities can be assigned one way or another, so that we can get on with the gaming, which is the most fun. But of what practical use are results based on arbitrary assumptions?*

In any matter of complexity, it is to be expected that different people will rate the possible outcomes differently. Where one analyst sees a prisoner's dilemma, another sees a zero-sum game with a saddle point, and still another sees a game requiring a mixed strategy. All may come to different conclusions, and all may be applying game theory correctly!

ARE PEOPLE RATIONAL?

Well over nine tenths of the applications of game theory purport to prescribe or predict human behavior. But game theory is not very good at predicting what people will do. That failing is tough to dismiss. The prescriptions of game theory are founded on the assumption of "rational" play. These prescriptions may not be the best when the other players are irrational.

The problem is something like a retailer's "bait and switch" tactic. You go to a car dealership because it advertises a car you want for $10,000, which is the lowest advertised price for that particular car. When you get to the dealership, the salesperson tells you that they're all out of that model. They do have another model for $12,000. The trouble is, you have no way of knowing if $12,000 is the best price for the other model, or even if you want that model. The only reason you came to that dealership was to get the advertised car. Now that they don't have it, you have to wonder if you might not have been better off going to a different dealer.

In game theory you generally commit to a strategy on the basis of a single potential outcome (a maximin or Nash equilibrium). If your opponent doesn't do as game theory advocates, you may find that you could have done better with a different strategy.

One of the first experimental challenges of game theory was a set of studies done at the RAND Corporation in 1952 and 1954. The research team, which included John Nash, tried to establish or refute the applicability of von Neumann's n-person game theory.

In the RAND experiments, four to seven people sat around a table. They played a "game" mimicking the general n-person game of von Neumann's theory. The subjects were told they could win money by forming coalitions. An umpire revealed the amount of cash that would be awarded to each possible coalition. Coalition members were allowed to split winnings among themselves in any way that seemed reasonable. The results were a confusing mess that owed less to *Theory of Games and Economic Behavior* than to *Lord of the Flies*. The RAND report (quoted in *Games and Decisions)* stated:

> *Personality differences between the players were everywhere in evidence. The tendency of a player to get into coalitions seemed to have a high correlation with talkativeness. Frequently, when a coalition formed, its most aggressive member took charge of future bargaining. In many cases, aggressiveness played a role even in the first formation of a coalition; and who yelled first and loudest after the umpire said "go" made a difference in the outcome.*
>
> *In the four-person games, it seemed that the geometrical arrangement of the players around the table had no effect on the result; but in the five-person game, and especially in the seven-person game, it became quite important. . . . In general as the number of players increased, the atmosphere became more confused, more hectic, more competitive, and less pleasant to the subjects. . . .*
>
> *It is extremely difficult to tell whether or not the observed results corroborate the von Neumann-Morgenstern theory. This is partly so because it is not quite clear what the theory asserts.*

That the frenetic subjects in the RAND experiment did not act as in von Neumann and Morgenstern's analysis is no indictment of the mathematical theory. However, the experiment served further notice, to any who might have still needed it, that game theory was not a good predictor of human behavior. These results must have been particularly discouraging to those hoping that game theory would quickly revolutionize economics. An economic theory must predict what flesh-and-blood people will do, whether those actions are rational or not.

One need not study multiperson games to find evidence of irrationality. Even more puzzling were experiments on the prisoner's dilemma. Like the Flood-Dresher experiment, most of these studies concerned *iterated prisoner's dilemmas.* The latter is a series of prisoner's dilemmas in which each player knows he will be interacting repeatedly with the other.

The iterated prisoner's dilemma has become such a popular subject for psychological studies that political scientist Robert Axelrod dubbed it "the *E. coli* of social psychology."[1] Anatol Rapoport estimated that 200 experiments concerning the prisoner's dilemma were published between 1965 and 1971.

THE OHIO STATE STUDIES

In the late 1950s and early 1960s, the Air Force sponsored a series of psychological studies of the prisoner's dilemma at Ohio State University. The results, published in a series of papers in the *Journal of Conflict Resolution,* offer little comfort to the believer in human rationality.

With some variations, the Ohio State experiments (and similar ones done elsewhere) worked like this. The subjects were pairs of students in an introductory psychology course, strangers to one another. They sat in carrels that prevented the subjects from seeing each other during the experiment. In front of each subject were two buttons, one red and one black. A researcher told the subjects that they would be asked to choose one of the buttons and push it, and would receive payoffs in pennies based on what both did. A table of payoffs was posted where each subject could consult it throughout the experiment.

None of this explanation used game-theory jargon. For our pur-

1. *E. coli* is a common bacterium used in experiments, the "guinea pig" of microbiology.

poses, it is enough to know that the red button was for defection and the black for cooperation. One of the payoff tables used was for a typical prisoner's dilemma—on the cheap:

	Black	Red
Black	3¢, 3¢	0, 5¢
Red	5¢, 0	1¢, 1¢

The buttons illuminated lights on a display panel visible to the researcher. From this information, he doled out the pennies. Pairs of subjects played each other a fixed number of times, usually fifty. After each payoff, subjects could determine which button the other subject had chosen from the size of the payoff. Each experiment was run on a number of different pairs of subjects to get a statistical sample.

The first study (by Alvin Scodel, J. Sayer Minas, Philburn Ratoosh, and Milton Lipetz) reported that most subjects defected most of the time. With the payoff table used, mutually cooperating subjects could have won three times as much money as they would have by defecting. Only two of the twenty-two pairs of subjects did that. Everyone else defected most of the time. (This result was different from that of Flood and Dresher's 1950 experiment. But the informal Flood-Dresher experiment followed just the one pair of subjects, not a statistically meaningful sample.)

In another version of the Scodel-Minas experiment, the subjects were permitted a "huddle." After twenty-five trials, they were allowed to discuss the game between themselves for two minutes. They were free to make a pact to push the black buttons—or even threaten to retaliate for red choices. That didn't happen. According to the researchers, ". . . the subjects appeared reluctant to arrive at a joint strategy. Remarks were aimed at the other player with the seeming intent of discovering what his strategy was going to be." The huddle had little effect in most pairs' play of the game.

Why were the subjects so reluctant to cooperate? The researchers put that very question to the subjects after the experiment. Typical answers were, "How did I know he would play black?" and "I just kept hoping he would play black when I pushed red."

According to the researchers, the subjects were more interested in *beating* their opponents' scores than in maximizing their own scores.

The researchers speculated that "a kind of culturally imposed norm leads people who are strangers to each other to act guardedly. It is better to assure one's self of at least equal footing with the other person than to run the risk of being bested by him."

J. Sayer Minas, Alvin Scodel, David Marlowe, and Harve Rawson tried to minimize the sense of competition by repeating the experiment, this time carefully avoiding any "competitive" words. "Game," "play," "win," and "lose" were stricken from the explanation of the experiment to the subjects. This did not much change things. In fact, the researchers reported that mutual cooperation never occurred more frequently than it would have if the players pushed black and red buttons at random!

This was nothing in the catalogue of irrationality. The researchers tried not only the prisoner's dilemma but other games in which the incentive to defect was nonexistent or negative. One game looked like this:

	Black	Red
Black	4¢, 4¢	1¢, 3¢
Red	3¢, 1¢	0, 0

Now this isn't even an interesting game. There's no reason to defect at all. No matter what, you're penalized a penny for pushing the red button. But the subjects *did* push the red button, 47 percent of the time.

Here defection must derive from a competitive impulse. Players who always cooperate rack up the maximum scores possible—but the game is a "tie." By defecting with a cooperating partner, a player wins less but increases his score *relative* to his opponent.

This finding is probably a very telling one. People learn how to play games not from the prisoner's dilemma but from ticktacktoe, bridge, checkers, chess, Trivial Pursuit, Scrabble—all of which are zero-sum games. They are zero-sum games because all they have to offer players is the psychological reward of being the winner, a reward that comes at the expense of the losers. This is true even of some games that offer the illusion of being non-zero-sum. In Monopoly, you acquire "real estate" and "cash"—but at the end of the game, it's just Monopoly money and the only thing that matters is who wins.

It's been speculated that this bent toward zero-sum games reflects the innately competitive nature of our society. I'm not so sure that it's not just the practical difficulty of packaging a game with stakes that people really care about. The Ohio State subjects did not take the games seriously because of the meager rewards. Pennies practically *are* Monopoly money. A truer non-zero-sum game is a TV game show where, backed by network ad revenues, the payoffs are cars, vacations, or thousands of dollars in cash. Then players pay more attention to increasing their own winnings and less to how the others are doing.

The Ohio State group wrote:

> *In our games, for example, the necessity to avoid an ego-deflating experience that could result from attempted collaboration that is not reciprocated could very well account for the prevalence of red plays. In [the prisoner's dilemma] this need to maintain self-esteem so dominated the monetary values in the matrix that subjectively players are not really in a dilemma. The choice is between doing as well as or better than the other person and running the risk of doing worse. Intuitively, it would seem that most people in our culture have the kind of self-image that necessitates the first choice even if it involves a deliberate sacrifice of something else . . .*

Players in the Ohio State studies weren't good at detecting tricks played on them. Part of one experiment was a hoax: a real subject was paired with a stooge. Whenever the real subject cooperated, he was told that his opponent had cooperated (and both won the reward payoff). Whenever a subject defected, he was told that his partner had defected (giving both the punishment). Though there were fifty trials in the experiment, not a single subject realized what was happening. When questioned later, all supposed that the matching replies were coincidence.

Dozens of variables, some verging on the bizarre, have been tested to see what effect they have on the rate of cooperation. Some published results are contradictory. Many studies have been rudimentary, with a smallish number of not necessarily typical subjects. Not all provocative findings have been followed up.

One of the Ohio State studies found no difference in cooperation rates among the sexes. Other studies have claimed that women cooperate more than men in the prisoner's dilemma, and still others say men become "protective" and are more inclined to cooperate when they know their partner is a woman. Scodel wondered if intelligent people acted differently in a prisoner's dilemma. He ran an informal test with some mathematics graduate students (he guessed they were smarter than the psychology undergraduates!) and found no meaningful difference in cooperation.

Anatol Rapoport found that doctors, architects, and college students cooperate more than small-business owners. A study at Miami University showed that when subjects could "punish" each other with electrical shocks, the cooperation rate was higher. I don't suppose that's surprising. Allen H. Stix claimed that Librium, the widely prescribed tranquilizer, increased cooperation and the amount of money won. Other studies purported to show that a barbiturate had little effect and amphetamines decreased cooperation and money won. Stix noted with some alarm reports that President Kennedy had taken amphetamines to remain alert during the 1961 summit meeting with Khrushchev.

About the only believable conclusion of these studies is that those inclined to cooperate in one context usually do so in other contexts. Some people are habitual cooperators, and some are habitual defectors. For instance, in another of the Air Force studies at Ohio State, Daniel R. Lutzker found a strong correlation between "internationalism" and cooperation. He created a psychological test that measured views on international cooperation. (One of the items held to measure cooperation was, "We should have a World Government with the power to make laws which would be binding to all its member-nations.") Then he ran the standard test. Lutzker's group of "isolationists" pushed the red button more than his "internationalists." Lutzker concluded bluntly that " 'Patriotism' and 'nationalism' appear to be related to a lack of trust in others and an incapacity to strive for mutual benefit even under conditions where cooperation leads to greater personal gain"—strong words, considering that Lutzker was a lieutenant in the psychology service at Fort Dix.

9

VON NEUMANN'S LAST YEARS

One guesses that there is a happy medium between the ivory tower and practicality. Toward the end of his life, von Neumann became consultant for a dizzying farrago of private and public concerns: defense agencies, the CIA, IBM, and Standard Oil, among others. Jacob Bronowski wrote flatly (1973) that von Neumann "wasted the last years of his life." He faults von Neumann's shift away from not only pure mathematics but pure science. Why von Neumann took on so many jobs that (seemingly) cannot have been the most intellectually satisfying is unclear. Possibly his father's emphasis on making money had a belated effect. Stanislaw Ulam wrote (1976) that von Neumann "seemed to admire generals and admirals and got along well with them. . . . I think he had a hidden admiration for people or organizations that could be tough and ruthless."

THE H-BOMB

One of the projects that occupied von Neumann's last years was the H-bomb. The necessary calculations were even more daunting than those needed for the fission bomb. In 1951 von Neumann helped design a computer for the Los Alamos laboratory. At one point von Neumann told Ulam, "Probably in its execution we shall have to perform more elementary arithmetical steps than the total in all the computations performed by the human race heretofore." The conjecture prodded the two mathematicians to try to verify it. They concluded it was wrong: the combined number-crunching of all the world's schoolchildren through history did in fact exceed the number of computations needed for the hydrogen bomb.

Even so, the H-bomb calculations took six months. This was estimated to be the equivalent of several lifetimes of human labor. Lewis Strauss gave von Neumann much credit for beating the Soviets to the H-bomb.

In the course of this work, von Neumann became a vocal proponent of atomic testing, often challenging those scientists who warned of long-term effects of radiation and radioactive fallout. Von Neumann offered an analysis in a memorandum to Strauss quoted in *Men and Decisions* (1962):

> *The present vague fear and vague talk regarding the adverse world-wide effects of general radioactive contamination are all geared to the concept that any general damage to life must be excluded . . . Every worthwhile activity has a price, both in terms of certain damage and of potential damage—of risks—and the only relevant question is, whether the price is worth paying. . . .*
>
> *It is characteristic, that we willingly pay at the rate of 30,000–40,000 additional fatalities per year—about 2% of our total death rate!—for the advantages of individual transportation by automobile. . . .*
>
> *The really relevant point is: Is the price worth paying? For the U.S. it is. For another country, with no nuclear industry and a neutralistic attitude in world politics it may not be.*

As intense as was von Neumann's dislike of communism, he had no use for Senator Joseph McCarthy's anti-Communist campaigns. One of the charges brought against Oppenheimer was that he had opposed the hydrogen bomb program, a project close to von Neumann's heart. It says a good deal for von Neumann's integrity that he defended Oppenheimer so vigorously. He testified that Oppenheimer was both loyal and trustworthy, and got in a few bright comebacks at the expense of the prosecutors. At one point the prosecutors described a highly hypothetical situation, insinuating that it was similar to Oppenheimer's actions, and asked von Neumann if he would have behaved differently. Von Neumann realized the trap and answered, "You are telling me now to hypothesize that someone else acted badly and ask me whether I would have acted in the same way. But isn't this a question of when did you stop beating your wife?"

Von Neumann also helped design the SAGE (Semi-Automatic Ground Environment), the computer system designed to detect a Soviet nuclear attack. While work proceeded on SAGE, von Neumann worried that the Soviets might be gearing for a surprise attack just before the system was in place. Cuthbert Hurd of IBM recalls that von

Neumann made a study to see if one of IBM's existing computers could be programmed to detect a Soviet attack in the interim. Von Neumann dictated the computer code to Hurd, who wrote it down. At one point, Hurd caught von Neumann in a mistake, an event so rare that it stuck in Hurd's mind. The calculations indicated that the existing computer would not be practical for the task.

A VERY FINE TIGER

A bitterness, a profound pessimism, shadowed von Neumann's final years. Its roots ran deeper than a rocky marriage or a sense of talents squandered.

The early 1950s were an age of prosperity for the von Neumanns. In 1951, J. Robert Oppenheimer had raised von Neumann's institute salary to a handsome $18,000. Many of the outside positions paid well, too. Marina, nearly grown, was introduced lavishly to society at the Debutante's New Year's Ball.

Yet Johnny and Klara were unable to make peace between themselves for long. One letter from Johnny to his wife describes the von Neumanns' conflicts in terms that echo with the second-guessing of game theory. Starting as an apology for yet another fight, the letter becomes a plea for support and then a somber commentary on marital trust and betrayal. Johnny claims that past misunderstandings have caused Klara to fear him and that she acts out of fear rather than reason. Johnny, too, says he is afraid of things about Klara that he hopes are not real. It is a letter of great anguish.

Von Neumann's sense of hopelessness extended to the human race itself. He saw technology putting ever more power in the hands of individuals. The technology of war was an obvious example, but by no means the only one. Individuals could not be expected to look out for the common welfare, and thus technologies, unwisely used, tended to produce a set of escalating problems with no solution in sight. It was no longer possible to run away from problems, either. As von Neumann put it, "we've run out of real estate." Future wars and catastrophes would be worldwide, placing the human race's survival in doubt.

In November 1954, the editors of *Fortune* magazine invited von Neumann to lunch to talk over ideas for an article. After the meeting, editor Hedley Donovan made a rough outline of the article from von Neumann's comments and sent it to him: "In places my outline quotes you directly; in places it paraphrases you; in some places, perhaps, it puts words in your mouth. I do not really mean to do the latter of course," Donovan wrote. This outline, which is franker and gloomier than the final article, gives an idea of von Neumann's thoughts at the time:

> *The world in general, and the U.S. in particular, is riding a very fine tiger. Magnificent beast, superb claws, etc. But do we know how to dismount? You see this as a very unstable world and a very dangerous world for the next 25 years. (Perhaps for much longer, but it's only 25 years we've asked you to write about.) . . .*
>
> *The relatively small number of people who have the power to destroy so much of the world—and do it so cheaply, and so quickly. . . .*
>
> *But the long-term net of your outlook is pessimistic. Whether any of your pessimism is the result of your being a Central European by birth is something that $10,000 worth of psychoanalysis might or might not determine. In any case, you are pessimistic because you cannot yourself visualize any form of world organization capable of controlling the world's instability and ending the danger. And you haven't heard of anybody else who's visualized that organization.*

Donovan suggested titling the article "A Very Fine Tiger." It was published (as "Can We Survive Technology?") in the June 1955 issue. One somber analogy from the outline appears in the article: "For the kind of explosiveness that man will be able to contrive by 1980, the globe is dangerously small, its political units dangerously unstable. . . . Soon existing nations will be as unstable in war as a nation the size of Manhattan Island would have been in a conflict fought with the weapons of 1900."

THE COMMISSIONER

On October 23, 1954, President Eisenhower appointed von Neumann to the Atomic Energy Commission. The post paid $18,000 a year. Since commissioners were required to resign all other consultancies, it actually meant less money than von Neumann had been earning. He had mixed feelings. In his *Adventures of a Mathematician,* Ulam recalls:

> *Just after Johnny was offered the post of AEC Commissioner . . . we had a long conversation. He had profound reservations about his acceptance because of the ramification of the Oppenheimer Affair. . . . The decision to join the AEC had caused Johnny many sleepless nights, he said. . . . But he was flattered and proud that although foreign born he would be entrusted with a high governmental position of great potential influence in directing large areas of technology and science. He knew this could be an activity of great national importance.*

Von Neumann requested a leave of absence from the institute, and accepted.

The *New York Times* judged von Neumann's appointment "a useful gesture of conciliation toward a larger group of scientists who have been unhappy about the Oppenheimer verdict." *Time* magazine said, "People who should know say that Von Neumann is eminently qualified to sit across the atomic table from the Russians in the greatest game in the world." Asked for comment, Albert Einstein told the press, "Johnny has a very good mind."

The *New York Times* ran a brief article on von Neumann after his nomination (October 25, 1954). Von Neumann made the usual pleasantries about peacetime uses of atomic power. "But I am convinced," he told the *Times,* "it will take a long time before it could be applied economically—especially in this country, where power is so cheap." The article painted a picture of comfortable respectability:

> *The Hungarian-born mathematician and professor sat in his living room with his wife, Klara, and his huge dog, Inverse.*
>
> *Nodding toward the hundreds of classical records about the walls, he said he was "anti-musical" but that his wife was a music lover.*
>
> *Mrs. von Neumann also is a mathematician, but their 19-year-old daughter, Marina, a liberal arts student at Radcliffe College, has no interest in the subject . . .*[1]

1. A premature assessment. Marina graduated first in Radcliffe's 1956 class and became a well-known economist and vice president of General Motors.

Dr. von Neumann does not smoke and drinks "very moderately," he said. His wife revealed his "passion," as she called it, for "cookies, candies, chocolates and sweet things." They both like Hungarian goulash and wine.

Ironically, some objections to the von Neumann appointment were grounded on the fact that he had supported Oppenheimer and thus might be another dreamy academic soft on communism. An editorial in the *Pittsburgh Post-Gazette* called the appointment "a strange one. . . . there is nothing in his record to show that Dr. von Neumann has ever had any executive experience . . ." The *Post-Gazette* writer theorized that von Neumann had been selected either to placate Oppenheimer supporters or to swing some votes to the Republicans in New York State. (The White House described von Neumann as a political independent.)

Von Neumann's Senate confirmation hearings began January 10, 1955. He described himself candidly as "violently anti-Communist, and a good deal more militaristic than most." He mentioned that his closest relatives in socialist Hungary were "only cousins." His appointment was confirmed on March 14, 1955.

Von Neumann and family moved to Washington. They lived in a comfortable yellow house at 1529 Twenty-ninth Street N.W. in fashionable Georgetown. There they continued a brisk entertaining schedule in a huge living room with two fireplaces.

As a commissioner, von Neumann became a public figure. He was profiled in feature articles; he appeared on public affairs TV shows. Von Neumann received the crank letters that are the due of the celebrity scientist. People who had invented games of one sort or another wrote him, some wanting marketing advice. Someone claimed to have found the "pattern" in prime numbers and thought von Neumann might like to know about it (letter dated September 14, 1956, in Library of Congress archives).

In 1956 *Good Housekeeping* magazine ran an article on Klara von Neumann and her husband with the improbable title, "Married to a Man Who Believes the Mind Can Move the World." One of the stranger examples of 1950s women's magazine journalism, it is a dogged attempt to humanize a not entirely promising subject. "What's it like to suspect your husband of being the smartest man on earth?" the article asks. "When Klara von Neumann, a slender brunette of Washington, D.C., glances at her husband, a plump, cheerful man who was born in Hungary fifty-two years ago, the thought sometimes occurs to her that she may be married to the best brain in the world."

Klara told *Good Housekeeping:* "He has a very weak idea of the geography of the house, by the way. Once, in Princeton, I sent him to get me a glass of water; he came back after a while wanting to know where the glasses were kept. We had been in the house only seventeen years. . . . He has never touched a hammer or a screwdriver; he does nothing around the house. Except for fixing zippers. He can fix broken zippers with a touch."

Klara says of their scientific salon, "I listen and to me it is the most fascinating conversation in the world. Sometimes when I revisit places where I used to spend much time, such as Cannes, in France, I am happy for a day or two. Then the talk about food and clothes begins to seem strange to me, as if something important is missing."

The *Good Housekeeping* article verges on black humor with its description of one of von Neumann's projects: "Recently Dr. von Neumann became interested in probing the consequences of the suggestion that by spreading dyes on the Arctic and Antarctic ice fields we could cut down the amount of energy they reflect away from the earth and thus warm the earth the few degrees necessary to turn it into a demitropical planet. Early computations show that the world's major ice fields could be effectively stained at a cost approximating what was spent to build the world's railroads. From this he has gone on to the idea that it is now conceivably possible to wage a new form of warfare, climate war, in which one country can alter unfavorably the climate in the enemy country."

"The idea of retiring to a charming little house and garden somewhere would be sheer death to him," Klara said of Johnny. "He is complicated, and life is complicated with such a man. But it is intensely rewarding. I like the lucidity of the mathematical world and the mathematical feeling that there is only one right answer to a problem. I like the subject matter of our lives. We are like others, we have personal problems, but we talk about them for only a little while, then get on to better things. I am his sounding board. Others would do, but I happen to be the one, and it is fascination itself."

THE MOMENT OF HOPE

One of the issues facing the AEC and the Eisenhower administration was disarmament. East and West had come to appreciate the paradoxical fact that their nuclear arsenals provided little security. Both sides claimed the world would be better off without nuclear weapons. But to cooperate by disarming would be to run the risk of the other side defecting—secretly keeping its bombs somewhere. The latter was von Neumann's objection to pro-disarmament colleagues at a January 1954 meeting of the General Advisory Committee. According to the minutes of the meeting, von Neumann "pointed out that we believe the Russians are better equipped for clandestine operations than we are and that we would be competing, at a disadvantage, with them in a clandestine armament race."

Ever since January 1952, the UN's disarmament commission had been frozen in a standoff. Whenever a Western nation proposed a disarmament plan, the Soviet Union vetoed it. Then the Soviets came back with their own plan, and the Western powers found it unacceptable.

In 1955 French and British delegates to the UN unveiled a new disarmament proposal. It called for the elimination of all nuclear weapons, everywhere, enforced by an international body. The United States endorsed the plan. That left only the Soviets unconvinced. Negotiations dragged on. The U.S., British, and French delegates reasoned and wheedled the Soviets; they appealed to their reason and sense of duty to humanity.

On May 10, 1955, an unexpected thing happened. The Soviets agreed. They did this by presenting their own plan, which contained all the provisions of the Anglo-French plan. British statesman Philip Noel-Baker called this "the moment of hope."

The Western powers scrutinized the 1955 Soviet plan for catches or hidden clauses and found none. Two days later, the U.S. delegate announced, "We have been gratified to find that the concepts which we have put forward over a considerable length of time, and which we have repeated many times during this past two months, have been accepted in a large measure by the Soviet Union."

A long recess postponed work on a treaty until the end of August. In mid-summer, the "Big Four" powers met for a summit meeting in Geneva. On July 17, von Neumann appeared on an NBC TV special, "Meeting at the Summit." The show opened with Bob Hope, of all people, and included remote segments with Bertrand Russell, William Randolph Hearst, Jr., David Brinkley, and other notables. The producers asked von Neumann to pose in front of a studio set with a blackboard and say:

> *What you see here is a mathematical expression of the power of a nuclear reaction. It is a scientific fact and a political fact. There is great concern among people about the destructive power of nuclear weapons and there are those who think that it would have been better if they had never been made.*
>
> *But science and technology must be neutral. . . .*

The segment ended with von Neumann voicing his often-repeated public view of the bomb: "It is not up to the scientist to decide how to use or control what he has developed. That is a matter for all the people and their political leaders."

The Geneva meetings were encouraging and marked by two surprises. Soviet Premier Nikolai Bulganin announced that the Soviet army was cutting its troop strength by 45,000. He invited the Western nations to reciprocate.

Eisenhower had an even bigger surprise for the Soviets—and the British and French delegates. He sketched a visionary proposal to eliminate all military secrets between the United States and the Soviet Union. Eisenhower suggested that each nation turn over all its military secrets to the other: "give to each other a complete blueprint of our military establishments, from beginning to end, from one end of our countries to the other." Then Eisenhower proposed "to provide within our countries facilities for aerial photography to the other country—we provide to you the facilities for aerial reconnaissance, where you can make all the pictures you choose and take them to your own country to study, you to provide exactly the same facilities for us, and we to make these examinations." This "open skies" plan would allow each nation to convince itself that the other was not planning a surprise attack.

Eisenhower announced his plan and paused for the translators. A peal of mountain thunder crackled outside, and the lights went out. "I didn't mean to turn the lights out," joked Eisenhower weakly. The translators repeated his words in the darkened hall.

The Soviets had no answer. At a buffet the next day Bulganin told Eisenhower that his plan would not work because it would be so easy to camouflage bombers. Eisenhower responded, "If you think it is, please show us how."

In the weeks after the summit, neither East nor West responded to the other's proposals. The United States didn't cut its own troop strength. The Soviets didn't agree to an exchange of military secrets. The press in both nations accused the other's leaders of seeking a propaganda victory. It was pointed out that a cut in ground forces meant less to the Soviets than it would to the United States and, conversely, the United States had less stake in military secrets than the Soviets did.

On August 5, 1955, Eisenhower created the President's Special Committee on Disarmament. The AEC's delegate to this committee was John von Neumann. On September 6, Harold Stassen, the new U.S. delegate to the UN, announced disheartening news. The United States "put a reservation" on its support of the disarmament plan.

What happened was, and continues to be, the object of speculation. The obvious suspicion is that the Eisenhower administration had never intended to go along with a disarmament agreement but wanted to look like a peacemaker. The Administration may have been so confident that the Soviets would never agree that they felt they could bluff and not be called on it. Supporting this is the fact that Eisenhower later admitted for the Dulles Oral History, "We knew the Soviets wouldn't accept it [the Open Skies proposal]. We were sure of that." It should be noted that if the U.S. leadership truly preferred the status quo to mutual disarmament, the situation was *not* a prisoner's dilemma.

In March 1956, Stassen discussed the new U.S. stand on disarmament in a speech to the UN. It was the U.S. position that hydrogen bombs and the threat of their use "constituted an atomic shield against aggression." The bomb was "an important safeguard of peace" and "a powerful deterrent of war." In April 1957, Stassen admitted that the United States wasn't interested in disarmament anymore: "It is our view that if armaments, armed forces, and military expenditures are brought down to too low a level, then . . . instead of the prospects of peace being improved, the danger of war is increased." He admitted wistfully, "There was a time when there was in my country considerable thought of a very extreme form of control and inspection, and a very low level of armaments, armed forces, and military expenditure. We have concluded that that extreme form of control and inspection is not practical, feasible, or attainable."

ILLNESS

In the summer of 1955 von Neumann fell on a slippery floor in one of the corridors of power he frequented. He hurt his left shoulder, and the pain did not go away. In mid-July he checked into Bethesda Naval Hospital for a few days of tests. He was discharged "apparently in satisfactory condition," he wrote Strauss: ". . . there doesn't seem to be anything somatically wrong with me . . ."

The pain continued. Another examination at Bethesda lead to a strong suspicion of cancer. Thus as Klara wrote, "the pattern of our active and exciting life, centered around my husband's indefatigable and astounding mind, came to an abrupt stop."

Von Neumann had bone cancer. He checked into Boston's New England Deaconess Hospital for surgery. His physician, Dr. Shields Warren, conducted a biopsy and concluded that von Neumann had a secondary cancer. Cancer cells had entered the bloodstream and were spreading throughout the body. Further tests identified the prostate gland as the primary site of the cancer.

"The ensuing months were of alternating hope and despair; sometimes we were confident that the lesion in the shoulder was a single manifestation of the dread disease, not to recur for a long time, but then indefinable aches and pains that he suffered at times dashed our hopes for the future," Klara wrote.

It has been suggested that von Neumann's cancer resulted from the radiation he received as witness to the Operation Crossroads atomic tests on Bikini atoll nearly a decade earlier. A number of physicists associated with the bomb succumbed to cancer at relatively early ages. Fermi died of cancer in 1954 at the age fifty-three. Oppenheimer would die in 1967 at the age of sixty-two. The evidence is far from conclusive, though, and the fact remains that people with no unusual exposure to radiation die from cancer, too.[2] (Von Neumann's papers at

2. I checked to see if any other notables present at the Crossroads test had died of cancer. Unfortunately, most of the newspaper obituaries don't give a cause of death. Out of nine persons for whom I was able to find a cause of death, two died of cancer: Generals Joseph W. Stilwell and A. C. McAuliffe, both celebrated war heroes. Stilwell died of liver cancer on Oct. 12, 1946. That's presumably too soon to have been caused by the

the Library of Congress include a telegram of June 19, 1956: "All your friends at Eniwetok send greeting and good wishes for your speedy recovery." The islet was the site of H-bomb tests.)

According to the *Life* magazine obituary, von Neumann asked his doctor, "Now that this thing has come, how shall I spend the remainder of my life?" Dr. Warren replied, "Well, Johnny, I would stay with the commission as long as you feel up to it. But at the same time I would say that if you have any important scientific papers—anything further scientifically to say—I would get started on it right away."

Von Neumann consequently knew he was going to die for about a year and a half. He maintained a busy schedule of meetings with people from the ballistic missile committee, Los Alamos, Princeton, and Livermore, disclosing the gravity of his condition to few. His arm was in a sling. When people asked about it, he muttered something about a broken collarbone. In late November 1955 lesions were found on von Neumann's spine. One night while leaving a party with Klara, von Neumann complained that he was "uncertain" about walking. His doctors gave him a wheelchair, and by January 1956 he was confined to it permanently.

He soon needed more sleep than his accustomed four or five hours. He cut out drinking. The 1956 *Good Housekeeping* article—which does not mention that its subject is dying—has him drinking orange juice ("which he prefers to a cocktail") at a party. A hot-line phone by his bed connected him directly with his AEC office. A limousine took him and his wheelchair to commission meetings.[3]

People gathered that he was dying but did not say so. Typical is a letter from Lewis Strauss (January 19, 1956) in which he rejoices at

---

Crossroads radiation. McAuliffe died of leukemia in 1975 at the age of seventy-seven. Another test observer, General L. H. Brereton, died of a heart attack in 1967 nine days after having undergone abdominal surgery for an undisclosed ailment. About two deaths in nine from cancer is presently average in the United States, so the reported deaths provide no evidence of a higher than usual cancer rate.

3. Was the wheelchair-bound von Neumann a model for the title character of Stanley Kubrick's 1963 film *Dr. Strangelove or: How I Learned to Stop Worrying and Love the Bomb?* Strangelove, "Director of Weapons Research and Development," is confined to a wheelchair. He speaks of having commissioned a defense study from the "Bland Corporation." As is often the case with satire, a number of models have been suggested (especially Werner von Braun and Edward Teller), and there is no reason to think the character was based on any specific individual. The screenplay by Kubrick, Peter George, and Terry Southern was loosely based on George's novel *Red Alert.* Peter Sellers, who played Strangelove, said in interviews that he based his portrayal on observations of Henry Kissinger. The wavy hair and glasses resemble Kissinger, not the balding von Neumann.

predictions of von Neumann's immanent recovery. Princeton's Albert Tucker wrote (March 27, 1956) that he had heard of von Neumann's improved health from Oskar Morgenstern. As late as June 11, 1956, von Neumann's secretary, Marie Janinek, showed only slightly more candor in turning down an invitation from the Aspen Institute of Humanistic Studies. Unfortunately, she said, von Neumann's recovery had not been as speedy as had been hoped.

In early 1956 von Neumann explored "retirement" plans. He put out feelers to MIT, Yale, UCLA, and the RAND Corporation for a job after leaving the AEC. It is evidence of the esteem in which von Neumann was held that these institutions were anxious to recruit him, even though they were aware he was seriously ill and might not have much time left. Von Neumann inquired about their medical coverage and life insurance, and whether insurance was available without a medical examination. Dean Paul A. Dodd of UCLA offered the care of some of the Medical Center's finest physicians, who had consented to treat von Neumann for free. Frank Collbohm of RAND offered $37,500 of life insurance without a medical examination and guarantee of a programming job for Klara for life. In March von Neumann accepted the UCLA offer. He never was well enough to start work.

One of von Neumann's last public appearances was in February 1956, when President Eisenhower presented him with the Medal of Freedom at the White House. Von Neumann in his wheelchair listened as the President read a flattering proclamation, then told Eisenhower, "I wish I could be around long enough to deserve this honor." Eisenhower could only reply, "Oh yes, you will be with us for a long time. We need you." The President pinned the red-ribboned medal on von Neumann's lapel.

Von Neumann had unfinished scientific business and set to work on it. He was interested in the brain as a model for future computers. He spoke of building a computer that would be like a very simple brain. This work lead to his being asked to give the 1956 Silliman Lectures on the computer and the brain at Yale University—an honor all the greater since neurology was not von Neumann's original field.

The Silliman lectures were scheduled for late March. Von Neumann wanted very much to give them as his valedictory to science. Klara recalled in the preface to *The Computer and the Brain,* "By March, however, all false hopes were gone, and there was no longer any question of Johnny being able to travel anywhere. Again Yale University, as helpful and understanding as ever, did not cancel the lectures, but suggested that if the manuscript could be delivered, someone else would read it for him. In spite of many efforts, Johnny could not finish writing his planned lectures in time; as a matter of tragic fate he could never finish writing them at all."

In April 1956 von Neumann checked into Walter Reed Hospital for what was to be a permanent stay. He set up office in his hospital suite. Secretary of Defense Charles Wilson and many of the Air Force top brass were among the stream of business visitors even at this late date. To help his friend, Strauss talked Eisenhower into giving von Neumann the Fermi Award for contributions to the field of atomic energy. Von Neumann became the first recipient after the namesake physicist. The award consisted of a gold medal and $50,000 cash—so much money in those days that the amount was halved for subsequent winners. Strauss delivered the medal to von Neumann in the hospital. "He insisted on keeping it within reach as long as he was conscious," Strauss wrote.

In a memorial address given in 1971, Strauss recalled:

> *On one dramatic occasion near the end, there was a meeting at Walter Reed Hospital where, gathered around his bedside and attentive to his last words of advice and wisdom, were the Secretary of Defense and his Deputies, the Secretaries of the Army, Navy and Air Force, and all the military Chiefs of Staff. The central figure was the young mathematician who but a few years before had come to the United States as an immigrant from Hungary. I have never witnessed a more dramatic scene or a more moving tribute to a great intelligence.*

For von Neumann, death came slowly. The *Life* magazine obituary includes the ominously tactful statement, "Von Neumann's body, which he had never given much thought to, went on serving him much longer than did his mind." The doctors discovered the cancer spreading further. They did not tell von Neumann. He was too sick to attend Marina's June wedding on Long Island. On July 3, the *New York Times* broke the news that von Neumann was "seriously ill" and had been for several months. Von Neumann managed a few meetings that summer. As late as July 13, he felt well enough to schedule a meeting with John McDonald, editor of *Fortune,* even though it was necessary to have von Neumann's brother Michael write the letter. The letter warned that von Neumann might be incoherent when they met.

Von Neumann's mother was a frequent visitor to his hospital room. In July 1956 she fell ill; two weeks later she was dead from cancer at the age of seventy-six. The family tried to conceal the death from von Neumann. Ever the second-guesser, von Neumann surmised the truth and fell into renewed grief.

The military did its best to keep von Neumann alive as long as possible. General J. W. Schwartz wrote to General Omar Bradley inquiring about an experimental cancer serum being tested at the UCLA Medical Center. Bradley replied that he would be willing to help in any way possible, in view of von Neumann's value to national defense. Bradley recommended against the use of the serum, though, saying it was too experimental.

Von Neumann suffered profound depression. At times he could discourse on mathematics or history, or remember conversations word for word from years ago; other times he would not even recognize family or friends. Wrote Heims, "Then came complete psychological breakdown; panic; screams of uncontrollable terror every night." In a short film made of von Neumann's life, Teller says, "I think that von Neumann suffered more when his mind would no longer function than I have ever seen any human being suffer." His illness was so devastating to his family, his brother Nicholas told me, that both he and Klara were moved to make out living wills forbidding the use of heroic means to keep them alive should they ever be in a similar situation.

Von Neumann's dementia seems not to have been an immediate effect of the cancer. According to Dr. Shields, the brain was unaffected except for excruciating pain. As the pain got worse, it was eased with mind-clouding drugs. This, combined with the recognition of his impending mortality, was responsible for his deteriorating mental state. Klara wrote in a letter to Abe Taub (October 23, 1956) that Johnny was receiving painkillers. She was grateful that the doctors were preventing him from suffering. Von Neumann's papers in the Library of Congress include an unsigned note describing one night of his illness (not in von Neumann's handwriting). The note records times of waking, sleeping, and taking a drug called Lotusate. Von Neumann complained about hiccoughs (the attendant did not notice any hiccoughs) and expressed vague worries about Air Force matters in the morning. The note concludes that this was a relatively tranquil night, with no confused talking in Hungarian. Von Neumann's orderlies were Air Force personnel, cleared for top-secret material in case he blurted military secrets.

The last von Neumann memory anecdote tells of his brother Michael reading him Goethe's *Faust* in the original German in the hospital. As Michael would pause to turn the page, von Neumann would rattle off the next few lines from memory.

DEATH

As the end neared, von Neumann converted to Catholicism, this time sincerely. *Life* reported, "One morning he said to Klara, 'I want to see a priest.' He added, 'But he will have to be a special kind of priest, one that will be intellectually compatible.' " A Benedictine monk, Father Anselm Strittmatter, was found to preside over his conversion and baptism. Von Neumann saw him regularly the last year of his life.

Of this deathbed conversion, Morgenstern told Heims, "He was of course completely agnostic all his life, and then he suddenly turned Catholic—it doesn't agree with anything whatsoever in his attitude, outlook and thinking when he was healthy." The conversion did not give von Neumann much peace. Until the end he remained terrified of death, Strittmatter recalled.

On February 1, 1957, N. F. Twining, Chief of Staff of the Air Force, wrote von Neumann, thanking him for his service on the Scientific Advisory Board, and asking him to accept a continuing appointment to the Board. Von Neumann died on February 8, 1957.

10

CHICKEN AND THE CUBAN MISSILE CRISIS

In late 1950 the Cambridge University Labour Club passed a resolution censoring its own president, Bertrand Russell. The resolution criticized Russell sharply for his advocacy of a nuclear war against the Soviet Union. Russell responded curtly, "I have never advocated a preventive war, as your members would know if they took the trouble to ascertain facts."

This was the first in a long string of denials that continued for most of the decade. In a letter published in the October 1953 issue of the *Nation,* Russell credited the whole thing to a Communist plot:

> *The story that I supported a preventive war against Russia is a Communist invention. I once spoke at a meeting at which only one reporter was present and he was a Communist, though reporting for orthodox newspapers. He seized on this opportunity, and in spite of my utmost efforts I have never been able to undo the harm. . . . "The New Statesman" in London wrote assuming the truth of the report, and it was only by visiting the editor in company with my lawyer that I induced "The New Statesman" to publish a long letter of refutation from me. You are at liberty to make any use you like of this letter, and I shall be glad if you can make its contents known to anybody who still believes the slanderous report.*

Why would Russell, one of the principal exponents of preventive war, repudiate it? Certainly, times were changing. The Soviet nuclear arsenal grew; the Americans got the H-bomb; the Soviets got the H-bomb. The nightmare was no longer being the victim of a surprise attack in which one's enemy would emerge victorious and unscathed. Now that both sides had second-strike capabilities, a nuclear war of any kind would be a general holocaust. Russell further must have felt uncomfortable with the paradox of having supported war, yet being active in pacifist and disarmament movements. In 1958 he became the first president of the Campaign for Nuclear Disarmament. He resigned two years later mainly because the organization wasn't militant enough for his tastes. In 1961 Russell was sent to prison for organizing a sit-down for nuclear disarmament.

In a 1959 BBC broadcast, Russell finally admitted his former stand on preventive war. Interviewer John Freeman asked, "Is it true or untrue that in recent years you advocated that a preventive war might be made against communism, against Soviet Russia?" Russell answered:

> *It's entirely true, and I don't repent of it. It was not inconsistent with what I think now. What I thought all along was that a nuclear war in which both sides had nuclear weapons would be an utter and absolute disaster. There was a time, just after the last war, when the Americans had a monopoly of nuclear weapons and offered to internationalize nuclear weapons by the Baruch proposal, and I thought this an extremely generous proposal on their part, one which it would be very desirable that the world should accept; not that I advocated a nuclear war, but I did think that great pressure should be put upon Russia to accept the Baruch proposal, and I did think that if they continued to refuse it might be necessary actually to go to war. At that time nuclear weapons existed only on one side, and therefore the odds were the Russians would have given way. I thought they would, and I think still that that could have prevented the existence of two equal powers with these means of destruction, which is what is causing the terrible risk now.*

What if push came to shove: would Russell really have favored bombing the Soviets, Freeman asked. "I should," Russell answered, adding that "you can't threaten unless you're prepared to have your bluff called."

Freeman asked why, then, Russell had repeatedly denied favoring preventive war. Russell said, "I had, in fact, completely forgotten that I had ever thought a policy of threat involving possible war desirable." There Freeman let the matter rest. It stretches belief, though, to see how Russell could have forgotten the substance of numerous speeches, letters, and articles written just months before the first denial.

CHICKEN

Our present concern with Russell is for his role in recognizing another game-theoretic dilemma, the game of "chicken." Like the prisoner's dilemma, chicken is an important model for a diverse range of human conflicts.

The adolescent dare game of chicken came to public attention in the 1955 movie *Rebel Without a Cause.* In the movie, spoiled Los Angeles teenagers drive stolen cars to a cliff and play a game they call a "chickie run." The game consists of two boys simultaneously driving their cars off the edge of the cliff, jumping out at the last possible moment. The boy who jumps out *first* is "chicken" and loses.

The plot has one driver's sleeve getting caught in the door handle. He plunges with his car into the ocean. The movie, and the game, got a lot of publicity in part because the star, James Dean, died in a hot-rodding incident shortly before the film's release. Dean killed himself and injured two passengers while driving on a public highway at an estimated speed of 100 mph.

For obvious reasons, chicken was never very popular—except in Hollywood. It became almost an obligatory scene of low-budget "juvenile delinquent" films in the years afterward. Wrote film critic Jim Morton (1986), "The number of subsequent films featuring variations on Chicken is staggering. Usually it was used as a device to get rid of the 'bad' kid—teens lost their lives driving over cliffs, running into trains, smacking into walls and colliding with each other. The creative abilities of Hollywood scriptwriters were sorely taxed as they struggled to think of new ways to destroy the youth of the nation."

Bertrand Russell saw in chicken a metaphor for the nuclear stalemate. His 1959 book, *Common Sense and Nuclear Warfare,* not only describes the game but offers mordant comments on those who play the geopolitical version of it. Incidentally, the game Russell describes is now considered the "canonical" chicken, at least in game theory, rather than the off-the-cliff version of the movie:

> *Since the nuclear stalemate became apparent, the Governments of East and West have adopted the policy which Mr. Dulles calls "brinkmanship." This is a policy adapted from a sport which, I am told, is practised by some youthful degenerates. This sport is*

> *called "Chicken!" It is played by choosing a long straight road with a white line down the middle and starting two very fast cars towards each other from opposite ends. Each car is expected to keep the wheels of one side on the white line. As they approach each other, mutual destruction becomes more and more imminent. If one of them swerves from the white line before the other, the other, as he passes, shouts "Chicken!" and the one who has swerved becomes an object of contempt. . . .*
>
> *As played by irresponsible boys, this game is considered decadent and immoral, though only the lives of the players are risked. But when the game is played by eminent statesmen, who risk not only their own lives but those of many hundreds of millions of human beings, it is thought on both sides that the statesmen on one side are displaying a high degree of wisdom and courage, and only the statesmen on the other side are reprehensible. This, of course, is absurd. Both are to blame for playing such an incredibly dangerous game. The game may be played without misfortune a few times, but sooner or later it will come to be felt that loss of face is more dreadful than nuclear annihilation. The moment will come when neither side can face the derisive cry of "Chicken!" from the other side. When that moment is come, the statesmen of both sides will plunge the world into destruction.*

Russell, of course, is being facetious in implying that Dulles's "brinkmanship" was consciously adapted from highway chicken. Herman Kahn's *On Thermonuclear War* (1960) credits Russell as the source of the chicken analogy.

Chicken readily translates into an abstract game. Strictly speaking, game theory's chicken dilemma occurs at the last possible moment of a game of highway chicken. Each driver has calculated his reaction time and his car's turning radius (which are assumed identical for both cars and both drivers); there comes a moment of truth in which each must decide whether or not to swerve. This decision is irrevocable and must be made in ignorance of the other driver's decision. There is no time for one driver's last-minute decision to influence the other driver's decision. In its simultaneous, life or death simplicity, chicken is one of the purest examples of von Neumann's concept of a game.

The way players rank outcomes in highway chicken is obvious. The worst thing that can happen is for both players *not* to swerve. Then—BAM!!—the coroner picks both out of a Corvette dashboard.

The best thing that can happen, the real point of the game, is to show your machismo by not swerving and letting the other driver swerve. You survive to gloat, and the other guy is "chicken."

Being chicken is the next to worst outcome, but still better than dying.

There is a cooperative outcome in chicken. It's not so bad if both players swerve. Both come out alive, and no one can call the other a chicken. The payoff table might look like the following. The numbers represent arbitrary points, starting with 0 for the worst outcome, 1 for the next-to-worst outcome, and so on.

	Swerve	Drive straight
Swerve	2, 2	**1, 3**
Drive straight	**3, 1**	0, 0

How does chicken differ from the prisoner's dilemma? Mutual defection (the crash when both players drive straight) is the most feared outcome in chicken. In the prisoner's dilemma, cooperation while the other player defects (being the sucker) is the worst outcome.

The players of a prisoner's dilemma are better off defecting, no matter what the other does. One is inclined to view the other player's decision as a given (possibly the other prisoner has already spilled his guts, and the police are withholding this information). Then the question becomes, why not take the course that is guaranteed to produce the higher payoff?

This train of thought is less compelling in chicken. The player of chicken has a big stake in guessing what the other player is going to do. A curious feature of chicken is that both players want to do the *opposite* of whatever the other is going to do. If you knew with certainty that your opponent was going to swerve, you would want to drive straight. And if you knew he was going to drive straight, you would want to swerve—better chicken than dead. When both players want to be contrary, how do you decide?

The game of chicken has *two* Nash equilibriums (boldface, lower left and upper right cells). This is another case where the Nash theory leaves something to be desired. You don't want two solutions, any more than you want two heads. The equilibrium points are the cases where one player swerves and the other doesn't (lower left and upper right).

Take the upper right cell. Put yourself in the place of the row player, who swerves. You're the chicken (1 point). Do you have any regrets? Well, looking at the big picture, you wish you had driven straight and the other player had been the chicken (3 points for you). But the first rule of Monday-morning quarterbacking is that you can regret your strategy only, not your opponent's. You wouldn't want to have driven straight while the other player did. Then both of you would have crashed (0 points). Given the other player's choice, your choice was for the best.

But look: The outcome where you drive straight and the other swerves is also an equilibrium point. What actually happens when this game is played? It's hard to say. Under Nash's theory, either of the two of the equilibrium points is an equally "rational" outcome. Each player is hoping for a different equilibrium point, and unfortunately the outcome may not be an equilibrium point at all. Each player can choose to drive straight—on grounds that it is consistent with a rational, Nash-equilibrium solution—and rationally crash.

Consider these variations:

(a) As you speed toward possible doom, you are informed that the approaching driver is *your long-lost identical twin.* Neither of you suspected the other's existence, but you are quickly briefed that both of you dress alike—are Cubs fans—have a rottweiler named Max. And hey, look at the car coming toward you—*another* 1957 firecracker-red convertible. Evidently, the twin thinks exactly the way you do. Does this change things?

(b) This time you are a perfectly logical being (whatever that is) and so is the other driver. There is only one "logical" thing to do in a chicken dilemma. Neither of you is capable of being mistaken about what to do.

(c) There is no other driver; it's a big mirror placed across the highway. If you don't swerve you smash into the mirror and die.

All these cases stack the deck in favor of swerving. Provided the other driver is almost certain to do whatever you do, that's the better strategy. Of course, there is no such guarantee in general.

Strangely enough, an *irrational* player has the upper hand in chicken. Take these variations:

(d) The other driver is suicidal and *wants* to die.

(e) The other driver is a remote-controlled dummy whose choice is made randomly. There is a 50 percent chance he will swerve and a 50 percent chance he will drive straight.

The suicidal driver evidently can be counted on to drive straight (the possibly fatal strategy). You'd rationally have to swerve. The random driver illustrates another difference between chicken and the prisoner's dilemma. With an opponent you can't second-guess, you might be inclined to play it safe and swerve. Of the two strategies in chicken, swerving (cooperation) has the maximum minimum. In the prisoner's dilemma, defection is safer.

VOLUNTEER'S DILEMMA

Just as the prisoner's dilemma has a multiperson version, so does chicken: the "volunteer's dilemma." You're sitting at home one night and the lights go out. You look outside and see that all the lights in the neighborhood are out. The electric company will send someone out to fix the problem—provided, that is, someone calls to tell them about it. Should *you* call? Naw—let someone else do it.

In a volunteer's dilemma, someone has to take on a chore that will benefit everyone. It doesn't matter who does it—but *everyone*'s in trouble if *no one* does it.

In two-person chicken, it is desirable that one player "volunteer" to swerve for the common good. When no one volunteers, both get stung. When both volunteer, each player kicks himself for not driving straight. The *n*-person version looks something like this:

	At least one person volunteers	**Everyone says, let someone else do it**
You volunteer	1	—
You say, let someone else do it	2	0

The "1" in the upper left cell signifies that you get the benefit of having the electric company notified (2) minus the slight inconvenience of phoning them. In fact, you might wonder what the fuss is about in the above anecdote: Why not just phone the power company and be done with it?

Let's adjust the payoffs. If the phone lines were also out and you had to hike three miles in the snow to notify the electric company, the spread between volunteering and not volunteering would be greater. Then you'd be more inclined to let someone else notify the electric company. You'd also be more worried that no one else would do it.

You're at a really strict boarding school. To defy the headmaster, all the students get together and steal the old school bell from the belfry. The next day, the headmaster is furious. He calls everyone into the auditorium and makes an offer: Provided that someone informs him of the whereabouts of the bell by the end of the day, that person or persons (who is clearly guilty) will receive a failing grade for the semester. If no one tells him where the bell is, *everyone* will receive a failing grade for the whole year. The students know that everyone is equally guilty, and all know where the bell is. Even the scapegoat(s) is better off than he would be if no one confessed. Do you volunteer?

Let's turn up the juice a little more. The worst form of the volunteer's dilemma occurs when the volunteer's payoff is almost identical with the "catastrophe" payoff when no one volunteers. Then there is a lifeboat situation where one person must "sacrifice" himself or all are in trouble—*but* with the all-important difference that the persons involved cannot draw straws or otherwise confer with each other.

This is similar to one of the dilemmas mentioned in the first chapter. You and ninety-nine friends are held captive in a problem box. Every person is in a separate soundproof cubicle. Each cubicle has a button. If you push the button you die. But if no one pushes the button before the big clock of doom on the wall strikes twelve, everyone dies.

The worst possible outcome is for no one to push the button. To you, the next-to-worst outcome is for you to push the button. Then you die a hero. Unfortunately, there is no guarantee that your death was necessary (someone else might push the button too) or even that it did any good (it is barely possible that *everyone* pushed the button and they all died anyway). The most desired outcome, of course, is for you to survive by having someone other than you push the button.

The volunteer's dilemma is widespread. The classic example in American urban lore is the 1964 murder of Kitty Genovese, the New York woman who was stabbed to death in the courtyard of her Kew Gardens apartment house while thirty-eight neighbors watched and heard her cries but did not come to her aid. Game theorist Anatol Rapoport noted (1988), "In the U.S. Infantry Manual published during World War II, the soldier was told what to do if a live grenade fell into the trench where he and others were sitting: to wrap himself around the grenade so as to at least save the others. (If no one "volunteered," all would be killed, and there were only a few seconds to decide who would be the hero.)" Another military example occurs in Joseph Heller's war novel *Catch-22.* When Yossarian balks at flying suicide missions, his superiors ask "What if everybody felt that way?" Yossarian responds, "Then I'd certainly be a damned fool to feel any other way. Wouldn't I?"

Rapoport notes that the Fuegian language of the natives of Tierra del Fuego contains the word *mamihlapinatapai,* meaning, "looking at each other hoping that either will offer to do something that both parties desire but are unwilling to do."

VOLUNTEER'S DILEMMA EXPERIMENTS

It is possible to play the volunteer's dilemma as a party game. Pass out slips of paper to a group and tell everyone to write either "$1.00" or "10 cents" on the paper. Provided at least one person writes "10 cents," everyone gets the amount he wrote. If everyone writes "$1.00," no one wins anything.

A volunteer's dilemma experiment was announced in the October 1984 issue of *Science 84* magazine. Readers were invited to send in cards asking for $20 or $100. The editors' original idea was that everyone would win what he asked for provided no more than 20 percent of the entries asked for $100.

In an offer like this, everyone can and should enter, asking for $20 at least. If everyone cooperates by asking for $20, everyone wins $20. But there is also room for some people to be greedy and ask for $100. As long as not too many people are greedy, each defector wins $100 and in no way hurts anyone else. The catch is that if too many people are greedy, no one gets anything.

A troubling thing about this and other volunteer's dilemmas with many participants is that individual defectors have little to feel guilty about. With thousands of participants, the chance that one's defection will be the one that puts the fraction of defectors over the 20 percent threshold is remote. Chances are either that the percentage of defectors will be well under 20 percent—in which case one's defection doesn't hurt anyone—or it will be well over 20 percent, in which case no one was going to win anything anyway. Of course, if everyone thought that way . . .

The magazine's publisher, the American Association for the Advancement of Science, chickened out over the money offer—though not before trying unsuccessfully to get Lloyd's of London to insure against a payout. Staff writer William F. Allman offered to put his future salary on the line as "collateral," but to no avail. The publishers stipulated that no money would be awarded; readers were merely asked to act *as if* the offer was as stated.

The magazine received 33,511 entries—21,753 people asked for $20, and 11,758 asked for $100. The defection rate was 35 percent, meaning that the publishers could have safely offered the money. It's hard to say if this defection rate is typical. The contest was announced in connection with an article on cooperation, which might have predisposed people toward cooperating. The craven refusal to offer real money might have been a factor, too. Isaac Asimov wrote in: "A reader is asked to check off $20 and consider himself a 'nice guy,' or check off $100 and consider himself *not* a nice guy. In such a case, everyone is going to take that option of pinning the nice guy label on himself, *since it costs no money to do so.*"

A lot of people enter contests without paying much attention to the rules, and there may have been some "naive" entrants asking for $100 in addition to the "premeditated" defectors. One defector quoted Blanche DuBois: "I've always depended on the kindness of strangers."

A total of $1,610,860 was requested. Had everyone who entered asked for $20, the payout would have been $670,220. The maximum that 33,511 people *could* have won, assuming that just under 20 percent asked for $100, is $1,206,380.

THE CUBAN MISSILE CRISIS

The Kennedy administration was receptive to the RAND Corporation's circle of strategists. Herman Kahn and Daniel Ellsberg (who later came to public attention for his role in releasing the Pentagon Papers) championed the notion that U.S.-Soviet conflicts were chicken dilemmas.

Why? By the early 1960s, leaders of both nations agreed that a nuclear war was the worst possible outcome of any situation. Unfortunately, the horror of a third world war did not guarantee cooperation to avoid war. Just as in highway chicken, the player who is most reckless is apt to prevail.

Any time U.S. and Soviet interests conflicted, one or both sides was tempted to threaten war. Not that they wanted war: but if one side could convince the other that it was serious, the other side might back down in order to prevent global holocaust. As Nikita Khrushchev said of nuclear war (quoting a Russian folk saying), "It will be too late to cry over lost hair after your head is cut off." Even being exploited is better than war.

The only way to guarantee peace would be to let the more belligerent nation always have its way—which hardly seems fair or rational. As RAND's theorists realized, the problem is fundamental and goes far deeper than the personalities of Nikita Khrushchev or John Kennedy.

The Cuban missile crisis of October 1962 has become the classic instance of a political chicken dilemma. The United States and Soviet Union probably drew closer to the nuclear brink than they ever had. In one of the strangest incidents in the history of game theory, Bertrand Russell, at the age of ninety, found himself in the thick of the very type of dilemma he had named.

Cuba had been a source of growing concern to Americans since 1959, when Fidel Castro overthrew the Batista regime. Castro espoused Marxism and began accepting economic support from the Soviet Union. Fears of a Soviet foothold in the Western Hemisphere were realized—as close as ninety miles from the beaches of Florida.

Castro's seizure of power sent many Cubans into exile in Florida. Some vowed to overthrow Castro. Kennedy's administration financed several plots. One, the disastrous 1961 Bay of Pigs invasion, lowered Kennedy's prestige among Americans. To Cubans, it did far more than that. Kennedy became an active enemy of the Cuban government.

In the summer of 1962, U.S. spy planes discovered nuclear missile bases under construction in Cuba. They were being built by the Soviets. The United States obviously wanted the missile bases removed. To get the Soviets to do that, it seemed that the United States had to threaten war—but no one really wanted a war.

The Soviets just as obviously wanted the missile bases to stay. To stand firm was to run the risk of war, which they didn't want, either. Each side hoped the other would back down. These facts were generally appreciated and commented upon at the time. *Newsweek* reported that Khrushchev had remarked to visiting poet Robert Frost "almost with contempt, that the United States was too 'liberal' to push any buttons in a showdown." After the crisis, the magazine noted that "The pull-back came almost the moment it was clear the U.S. was prepared to fight."

On October 22, Kennedy took the hard line. Knowing that more Soviet ships were approaching Cuba, Kennedy announced a naval blockade of the island. The announcement brought an equally stern reply from Khrushchev. Should American ships stop the Soviet ships from landing, "we would then be forced for our part to take all action that is necessary," Khrushchev said.

Russell's bizarre role in the crisis stems from the fact that he liked to get things off his chest by writing letters to the editors of newspapers and even to heads of state. In 1957 Russell had published open letters on disarmament to Eisenhower and Khrushchev in the *New Statesmen.* Khrushchev had responded right away. The Eisenhower letter brought a reply from John Foster Dulles two months later.

As tensions began to heighten over Cuba, Russell composed a statement expressing his views. He felt that the United States was in the wrong. Cuba had consented to having the Soviets build bases; what right did the United States have to say what could or could not be built in Cuba? The United States had bases in nations bordering the Soviet Union. Russell warned that the crisis could escalate to nuclear war. He sent this statement to the press on September 3, 1962.

This statement was ignored. Undeterred, Russell sent a telegram to U Thant on October 18. Russell asked if he could address the General Assembly of the UN. The UN responded with a polite no, citing protocol.

When Kennedy announced his blockade, Russell sprang into action. He produced and distributed an almost hysterical leaflet. It read:

Statement re: CUBA CRISIS

YOU ARE TO DIE	Not in the course of nature, but within a few weeks, and not you alone, but your family, your friends, and all the inhabitants of Britain, together with many hundreds of millions of innocent people elsewhere.
WHY?	Because rich Americans dislike the Government that Cubans prefer, and have

used part of their wealth to spread lies about it.

WHAT CAN YOU DO?

You can go out into the streets and into the market place, proclaiming, 'Do not yield to ferocious and insane murderers. Do not imagine that it is your duty to die when your Prime Minister and the President of the United States tell you to do so. Remember rather your duty to your family, your friends, your country, the world you live in, and that future world which, if you so choose, may be glorious, happy, and free.'

AND REMEMBER:

CONFORMITY MEANS DEATH
ONLY PROTEST GIVES A HOPE OF LIFE

Bertrand Russell
23rd October, 1962

Russell also sent five telegrams to world leaders, including Kennedy, Khrushchev, and U Thant. There was no sugar-coating in the cable to Kennedy. It read, "Your action is desperate. Threat to human survival. No conceivable justification. Civilized man condemns it. We will not have mass murder. Ultimatums mean war. . . . End this madness."

The telegram to Khrushchev was more sympathetic: "I appeal to you not to be provoked by the unjustifiable action of the United States in Cuba. The world will support caution. Urge condemnation to be sought through the United Nations. Precipitous action could mean annihilation for mankind."

Did these telegrams have any effect on the course of events? Russell thought they did. The following year, he published a book on his role in the crisis, *Unarmed Victory.* Most reviewers at the time didn't buy Russell's claims. The *Observer* snipped that "old man's vanity had been at work." To the *Spectator* it was an object of "contempt and pity." Russell gets short shrift in many histories of the crisis, including that by Robert Kennedy.

But however biased Russell's own account may be, his telegrams were read and acted upon. Khrushchev had Tass print a public reply to Russell's wire on October 24. The Soviet leader's long letter to Russell included assurances that:

. . . I understand your worry and anxiety. I should like to assure you that the Soviet government will not take any reckless decisions, will not permit itself to be provoked by the unwarranted actions of the United States of America. . . . We shall do everything in our power to prevent war from breaking out. . . . The question of war and peace is so vital that we should consider a top-level meeting in order to discuss all the problems which have arisen, to do everything to remove the danger of unleashing a thermonuclear war. As long as rocket nuclear weapons are not put into play it is still possible to avert war. When aggression is unleashed by the Americans such a meeting will already become impossible and useless.

After this, TV, radio, and newspaper interviewers descended on Russell's home in North Wales. The press described him as a "nonagenarian intellectual in carpet slippers." Some wondered if a man so aged should be poking about in international relations. Meanwhile, on October 26, Kennedy replied to Russell's telegram with peevish wit.

I am in receipt of your telegram. We are currently discussing the matter in the United Nations. While your messages are critical of the United States, they make no mention of your concern for the introduction of secret Soviet missiles into Cuba. I think your attention might well be directed to the burglars rather than to those who caught the burglars.

This reply was quickly picked up by the press, too. By then the Russell telegrams were being played almost for comic relief against the building crisis.

The media circus aside, Khrushchev's letter was encouraging. It was far more conciliatory than the Soviet leader's most recent letter to Kennedy. Khrushchev said nothing in the Russell letter about Cuba's, or the Soviet Union's, right to have missiles in Cuba. His main concern appeared to be saving face. The only Soviet demands mentioned were modest to the vanishing point: an end to the blockade and that America not launch its missiles.

In *A Thousand Days: John F. Kennedy in the White House,* Arthur M. Schlesinger, Jr., confirms that the Administration saw the reply to Russell as an attempt to save face. He says that on October 24:

I received a call from Averell Harriman. Speaking with usual urgency, he said that Khrushchev was desperately signaling a desire to cooperate in moving towards a peaceful solution. . . . Harriman set forth the evidence: Khrushchev's suggestion of a summit meeting in his reply to Bertrand Russell . . . the indications that afternoon that the nearest Soviet ships were slowing down and changing course. "This was not the behavior of a man who wanted war," Harriman said, "it was the behavior of a man begging our help to get off the hook . . . we must give him an out . . . if we can do this shrewdly we can downgrade the tough group in the Soviet Union which persuaded him to do all this. But if we deny him an out, then we will escalate this business into a nuclear war."

This makes it sound like Russell played a useful role. Khrushchev's reply to Russell may have been "merely" a face-saving ploy, but face-saving is precisely what is called for in a chicken dilemma. When one side can find a good excuse to back down, the dilemma ceases to be a dilemma. One reason the world seemingly came so close to war may have been that Kennedy felt he couldn't back down so soon after the humiliating Bay of Pigs affair.

It was easier for Khrushchev to take a softer line in a telegram to Russell, an intellectual respected in the East and West, than to his adversary, Kennedy. The question is whether Russell played an essential role. Khrushchev might have issued a public reply to someone else's letter, or sent out feelers some other way.

Nevertheless, those who felt Russell was meddling in affairs better left to professionals soon found proof of their claim. Russell unwittingly made a major blunder.

On October 26, Russell sent another telegram to Khrushchev, thanking him for his reply. This time, Russell sketched a possible settlement. He asked Khrushchev to make a "unilateral gesture"—dismantling the Cuban missile bases—and then to request that the Americans respond in kind. "The abandonment of the Warsaw Pact, for example, could be the basis for a request that the Americans make a similar gesture in Turkey and Iran, in West Germany and in Great Britain." Russell was suggesting that the Americans might remove their long-established missile bases from the nations mentioned. This off-the-top-of-his-head settlement was seemingly asking a lot more of America than of the Soviet Union.

Russell was in no position to negotiate for the Americans. If he had been, he would have known that a tentative settlement was already in the works at the time of the second telegram. The Soviet embassy had offered to remove the missiles from Cuba, and further vow never to place such weapons in Cuba in the future, if only the United States would remove its missiles from Florida under UN inspection.

This, of course, was a much better deal for the United States. The Florida missiles were mainly useful in attacking Cuba, and would be superfluous if there were no Soviet missiles in Cuba. The missiles Russell was talking about, in Europe and the Middle East, could be targeted at the Soviet Union, Europe, and much of Asia.

Unfortunately, on the twenty-seventh, Khrushchev changed his tune. Tass disseminated a letter offering to remove the Cuban bases if the United States removed its missile bases in *Turkey.*

Some accounts of the crisis claim that this was the first time the Soviets had mentioned Turkey—the first time anyone connected with the crisis had, except Russell. That's not quite true. On October 23, Soviet Defense Minister Rodion Y. Malinovsky told a Western diplomat in Rumania, "In Turkey, the Americans are holding a dagger at our chest. Why can't we hold a dagger at the American chest in Cuba?" This shows that at least one Soviet leader saw an analogy between Cuba and Turkey before Russell mentioned it. The *Red Star,* the Soviet Army's newspaper, suggested a Cuba-Turkey swap on the twenty-sixth. It is unclear if this was prompted by Russell's telegram sent the same day.

Kennedy quickly rejected the Turkey offer. The deal was off.

The next day, Russell transmitted yet another message to Khrushchev. He proposed another solution—in a nutshell, the resolution that was actually reached.

> *The U.S. rejection of your proposals to trade Soviet installations in Cuba against NATO installations in Turkey is totally unjustifiable and is a sign of insane paranoia. . . . It seems to me, therefore, that you ought to dismantle Soviet installations in Cuba under the guarantee of inspection by UNO, demanding only, in return, that when UNO gives the necessary guarantee, the American blockade of Cuba should be lifted. . . . I have not made [this proposal] public, and I shall not do so unless it meets with your approval. It asks of you a sacrifice which you may find excessive and intolerable.*

Late on the twenty-eighth, hours after Russell sent this telegram, Khrushchev offered to dismantle all Cuban missile bases and to remove Soviet emissaries from Cuba. He asked essentially nothing from the United States in return. Thus the crisis ended.

Russell's *Unarmed Victory* implies that the telegram induced the Soviet offer. "The overwhelming impression created is that the end of the crisis followed Russell's cable as effect follows cause," biographer Ronald Clark said skeptically (1978) of the Russell account.

On the other hand, Khrushchev and U Thant acknowledged Russell's role in letters or memoirs. U Thant wrote in his 1978 memoirs that "I felt at the time, and still feel, that Khrushchev's positive reply to my first appeal of October 24 was, at least in part, due to Earl Russell's repeated pleadings to him, and to his congratulating him on 'his courageous stand for sanity.' "

Russell scholar Al Seckel has suggested that Khrushchev found Russell particularly useful because of Russell's outspoken anti-American sentiments. U Thant refused to condemn the United States for its role in the Cuba crisis, as Khrushchev had wanted. Khrushchev knew that Russell was brimming with criticism for Kennedy's actions, and that Russell's views would be widely publicized if Khrushchev focused attention on him by responding to the telegram. Russell's past criticism of the Soviet Union lent a kind of credibility. Russell had proposed dropping atomic bombs on Moscow—no one could claim he was a puppet of the Kremlin.

Like all applications of game theory to foreign relations, the identification of the Cuban crisis with the game of chicken depends on a number of assumptions about what would have happened. The usual analysis glibly assumes that mutual intransigeance would have lead to nuclear war. This does not follow as the night the day. It is fairer to say that it would have lead to a chance of nuclear war, and even a slight chance of such a dreadful prospect must be taken seriously. Estimates of probability vary, though, and it is less clear that both sides would prefer loss of face to the chance of war.

In 1962 Robert F. Kennedy said, "We all agreed in the end that if the Russians were ready to go to nuclear war over Cuba, they were ready to go to nuclear war, and that was that. So we might as well have the showdown then as six months later."

As in much of foreign relations, motivations are often concealed by several layers of rhetoric. In 1987 Theodore Sorenson wrote of the same crisis, "The President drew the line precisely where he thought the Soviets were not and would not be; that is to say, if we had known that the Soviets were putting 40 missiles in Cuba, we might under this hypothesis have drawn the line at 100, and said with great fanfare that we would absolutely not tolerate the presence of more missiles in Cuba . . . one reason the line was drawn at zero was because we simply thought the Soviets weren't going to deploy any there anyway."

The missile crisis was not Russell's last moment in the limelight. Despite his icy exchange with Kennedy, it was Russell who headed the Who Killed Kennedy Committee after the release of the Warren Report. In the late 1960s, Russell lead protests against the Vietnam war. He died at home in 1970.

THE MADMAN THEORY

The most disturbing thing about the chicken dilemma is the "advantage" an irrational player has or seems to have. In *On Escalation,* Herman Kahn claimed, "Some teenagers utilize interesting tactics in playing 'chicken.' The 'skillful' player may get into the car quite drunk, throwing whisky bottles out the window to make it clear to everybody just how drunk he is. He wears very dark glasses so that it is obvious he cannot see much, if anything. As soon as the car reaches high speed, he takes the steering wheel and throws it out the window. If his opponent is watching, he has won. If his opponent is not watching, he has a problem . . ."

Stated Kahn, "It is clear from the above why many people would like to conduct international relations the way a teenager plays 'chicken.' They believe that if our decision-makers can only give the appearance of being drunk, blind, and without a steering wheel, they will 'win' in negotiations with the Soviets on crucial issues. I do not consider this a useful or responsible policy."

Kahn claimed that differences of state etiquette between East and West helped the Soviets in cold-war chicken dilemmas. Khrushchev, Kahn said, sometimes lost his temper in public and became, or pretended to become, irrational. U.S. Presidents aren't supposed to do that.

Even so, President Nixon apparently used such a strategy in the Vietnam war. In *The Ends of Power* (1978), presidential aide H. R. Haldeman wrote:

> *The threat was the key, and Nixon coined a phrase for his theory which I'm sure will bring smiles of delight to Nixon-haters everywhere. We were walking along a foggy beach after a long day of speechwriting. He said, "I call it the Madman Theory, Bob. I want the North Vietnamese to believe I've reached the point where I might do* anything *to stop the war. We'll just slip the word to them that, 'for God's sake, you know Nixon is obsessed about Communism. We can't restrain him when he's angry—and he has his hand on the nuclear button'—and Ho Chi Minh himself will be in Paris in two days begging for peace."*

A problem with the madman theory is that two can play at that game, confusing already complex matters. Kahn concludes,

> *"Chicken" would be a better analogy to escalation if it were played with two cars starting an unknown distance apart, traveling toward each other at unknown speeds, and on roads with several forks so that the opposing sides are not certain that they are even on the same road. Both drivers should be giving and receiving threats and promises while they approach each other, and tearful mothers and stern fathers should be lining the sides of the roads urging, respectively, caution and manliness.*

11

MORE ON SOCIAL DILEMMAS

We now have met two dilemmas of fundamental importance in human affairs. Are there others?

In 1966, Melvin J. Guyer and Anatol Rapoport, then both at the University of Michigan, catalogued all the simple games. The simplest true games have two players making a choice between two alternatives. It is reasonable to think that these "2 × 2" games ought to be particularly important and common. The prisoner's dilemma and chicken are 2 × 2 games, of course. There are seventy-eight distinct 2 × 2 games when payoffs are simply ranked rather than assigned numerical values.

A symmetric game is one where the payoffs are the same for each player under comparable circumstances—where neither player is in a privileged position. If player A is the lone cooperator, his payoff is the same as player B's would be if he was the lone cooperator, and so forth. Symmetric games are the easiest to understand, and probably the most important in social interactions. People are much the same the world over. No conflicts are so common or so bitter as those between people wanting the same thing. So let's look at the symmetric games.

There are just four payoffs to worry about in a symmetric 2 × 2 game. Let "CC" be the payoff, to each of the two players, when both cooperate. "DD" is the payoff when both defect. When one player cooperates and the other defects, the payoff to the lone cooperator will be called "CD," and the payoff to the lone defector, "DC."

All the variety of 2 × 2 symmetric games rests in the relative values of the four payoffs CC, DD, CD, and DC. As usual, let's have the players rank them in order of preference (and they must agree on the ranking, it being a symmetric game). Let's further assume that there are no "ties," that there is always a distinct preference between any two payoffs.

Any given preference ordering of the four payoffs defines a game. For instance, when

DC>CC>DD>CD

—meaning that the DC outcome is preferred to CC, which is preferred to DD, which is preferred to CD—the game is a prisoner's dilemma. (The usual further requirement, that the average of the DC and CD payoffs be less than the CC payoff, applies only when players have assigned numerical values to the payoffs. Here we're just ranking them.)

There are twenty-four possible rankings of four payoffs, and thus twenty-four symmetric 2×2 games. Not all are dilemmas. In most, the correct strategy is obvious.

The disturbing thing about the prisoner's dilemma and chicken is the way that the common good is subverted by individual rationality. Each player desires the other's cooperation, yet is tempted to defect himself.

Let's see what this means in general terms. The payoff CC must be preferred to CD. That means that you're better off when your partner returns your cooperation. Also, DC must be better than DD. You still hope the other player cooperates, even when you yourself defect.

Of the twenty-four possible orderings of four payoffs, exactly half have the CC payoff higher than the CD payoff. Likewise, exactly half have DC preferred to DD. Just six of the possible orderings simultaneously meet both requirements. The six look like this:

CC>CD>DC>DD
CC>DC>CD>DD
CC>DC>DD>CD
DC>CC>CD>DD
DC>CC>DD>CD
DC>DD>CC>CD

Not all of these are troublesome. If defection is bad all around, everyone will avoid it. For a true dilemma to exist, there must be a temptation to defect—otherwise, why defect?

In the prisoner's dilemma, there is every temptation to defect. No matter what the other player does, you are better off defecting. The temptation need not be that acute to pose a dilemma. You may have a hunch what the other player is going to do, and know it is in your advantage to defect provided the hunch is right. This could cause you to defect, even though there may be no incentive to defect if your hunch is wrong.

We require, then, that *either* of two conditions be met. Either there is an incentive to defect when the other player cooperates (DC>CC), or there is an incentive to defect when the other player defects (DD>CD) —or both.

This rules out two of the games above. With the payoffs ordered CC>CD>DC>DD or CC>DC>CD >DD, there is no incentive to defect at all. Not only is mutual cooperation the best possible outcome, but a player is guaranteed to do better by cooperating, no matter what the other player does.

Crossing these two games off the list leaves just four games. Each is important enough to rate a name.

DC>DD>CC>CD	Deadlock
DC>CC>DD>CD	Prisoner's Dilemma
DC>CC>CD>DD	Chicken
CC>DC>DD>CD	Stag Hunt

All four are common games in real-life interactions. For that reason they are called "social dilemmas." All are closely related, too. Each of the other three games can be derived from the prisoner's dilemma by switching two of the payoffs in the order of preference. You can think of the prisoner's dilemma as a center of gravity around which the others orbit. Chicken is a prisoner's dilemma with punishment and sucker payoffs reversed. The stag hunt is the prisoner's dilemma with the preferences of the reward and temptation payoffs switched. Deadlock is a prisoner's dilemma with the reward and punishment payoffs switched. Let's examine the two new games.

DEADLOCK

Deadlock is the least troublesome of the four. The game looks like this (where the worst outcome is defined to have a utility of 0):

	Cooperate	**Defect**
Cooperate	1, 1	0, 3
Defect	3, 0	**2, 2**

The deadlock player quickly surmises that he should defect. As in the prisoner's dilemma, a player does better defecting no matter what his partner does. The difference is that players actually prefer mutual defection to mutual cooperation.

Players want to defect in the hope of getting 3 points. But even if they both defect, it's no tragedy—each gets his second-best outcome. That's better than they could do with cooperation. Deadlock is not properly a dilemma at all. There is no reason for wavering: you should defect. Mutual defection is a Nash equilibrium.

Deadlock occurs when two parties fail to cooperate because neither really wants to—they just want the other guy to cooperate. Not all failures to come to arms control agreements are the result of prisoner's dilemmas. It may be that neither side truly wants to disarm. Possibly that was the case with the U.S.-Soviet "moment of hope" in 1955.

STAG HUNT

The stag hunt is more of a dilemma. Like chicken, it recalls dilemmas of adolescence. It's late May, next to the last day of school. You and a friend decide it would be a great joke to show up on the last day of school with some ridiculous haircut. Egged on by your clique, you both *swear* you'll get the haircut.

A night of indecision follows. As you anticipate your parents' and your teachers' reactions to the haircut, you start wondering if your friend is really going to go through with the plan.

Not that you don't want the plan to succeed: the best possible outcome would be for both of you to get the haircut.

The trouble is, it would be awful to be the *only* one to show up with the haircut. That would be the worst possible outcome.

You're not above enjoying your friend's embarrassment. If you *didn't* get the haircut, but the friend did, and looked like a real jerk, that would almost be as good as if both of you got the haircut.

After mulling things over, you conclude that it wouldn't really be so bad if *no one* got the haircut. Maybe everyone will just forget about it. (This is what your mother says will happen.)

Of the possible outcomes, your first choice is for mutual cooperation (both get the haircut), second is unilateral defection (you don't get the haircut, your friend does), third is mutual defection (both chicken out), and fourth is unilateral cooperation (you get the haircut, the friend doesn't). Assume that your friend has the same preferences. The barbershop at the mall closes at nine. What do you do?

The peculiar thing about the stag hunt is that it shouldn't be a dilemma at all. You should certainly cooperate—that is, get the haircut. If both of you do, both will get the best possible payoff. What spoils things is the possibility that the friend won't be so rational. If the friend chickens out, you want to chicken out, too.

The game of stag hunt has been known by many names in the literature of game theory, among them "trust dilemma," "assurance game," and "coordination game." These colorless terms have been firmly superseded by the more poetic "stag hunt," which derives from a metaphor in Swiss-born philosopher Jean-Jacques Rousseau's *A Discourse on Inequality* (1755).

The writings of Rousseau idealized primitive man and held that most social ills were the product of civilization itself. He based his philosophy on a speculative and inaccurate conception of prehistory. In *A Discourse on Inequality* Rousseau attempted to offer "scientific" support for his arguments with traveler's tales that have an almost magic-realist quality today. He tells of the orangutan presented to the Prince of Orange, Frederick Henry. It slept in a bed with its head on the pillow and could drink from a cup. From reports of orangutans forcing unwanted sexual attention on women, Rousseau speculates that they were the satyrs of Greek mythology.

Part Two of the *Discourse* theorizes that the first human societies began when people forged temporary alliances for hunting. The "stag" is a deer in Maurice Cranston's translation:

If it was a matter of hunting a deer, everyone well realized that he must remain faithfully at his post; but if a hare happened to pass within the reach of one of them, we cannot doubt that he would have gone off in pursuit of it without scruple and, having caught his own prey, he would have cared very little about having caused his companions to lose theirs.

The point is that no individual is strong enough to subdue a stag by himself. It takes only one hunter to catch a hare. Everyone prefers stag to hare, and hare to nothing at all (which is what the stag party will end up with if too many members run off chasing hares).

The payoffs, in arbitrary points, look like this:

	Hunt stag	Chase hare
Hunt stag	**3, 3**	0, 2
Chase hare	2, 0	1, 1

Obviously, mutual cooperation is a Nash equilibrium. The players can't do any better no matter what. Temptation to defect arises only when you believe that others will defect. For this reason the dilemma is most acute when one has reason to doubt the rationality of the other player, or in groups large enough that, given the vagaries of human nature, some defections are likely.

A mutiny may pose a stag hunt ("We'd all be better off if we got rid of Captain Bligh, but we'll be hung as mutineers if not enough crewmen go along.") Elected representatives sometimes favor a bill but are reluctant to vote for it unless they are sure it will pass. They don't want to be in the losing minority. This was apparently the case with some of the U.S. senators voting on President Bush's 1989 constitutional amendment to make burning of the U.S. flag a federal crime. Most opponents of the bill objected to it as a violation of freedom of expression. At the same time, they feared that if they voted against it and it passed, their opponents would brand them unpatriotic or "in favor of flag burning" in the next election. Senator Joseph R. Biden, Jr., an opponent of the bill, made the striking claim, "More than 45 senators would vote 'no' if they knew they were casting the deciding vote."

Arguably, the stag hunt describes the ethical dilemma of the scientists who built the atomic bomb. Roughly: The world would be better off without the bomb ("I personally hope very much that the bombs will not explode, no matter how much effort is put into the project," Harold Urey said of the hydrogen bomb in his 1950 speech). But we have to try to build it because our enemy will. Better we have the bomb than our enemy; better both sides have the bomb than just our enemy.

In the wake of professional hockey player Teddy Green's 1969 head injury, *Newsweek* stated:

> *Players will not adopt helmets by individual choice for several reasons. Chicago star Bobby Hull cites the simplest factor: "Vanity." But many players honestly believe that helmets will cut their efficiency and put them at a disadvantage, and others fear the ridicule of opponents. The use of helmets will spread only through fear caused by injuries like Green's—or through a rule making them mandatory . . . One player summed up the feelings of many: "It's foolish not to wear a helmet. But I don't—because the other guys don't. I know that's silly, but most of the players feel the same way. If the league made us do it, though, we'd all wear them and nobody would mind."*

ASYMMETRIC GAMES

The social dilemmas described above are symmetric games in which both players share the same preferences. Preferences need not match, however. It is possible for one player's preferences to be those of a prisoner's dilemma and the other player's preferences to be those of chicken or stag hunt or something else. Some of these hybrid games have been suggested as models for human conflicts.

The game of bully is a cross between chicken and deadlock. One player has the same preferences as a chicken player. He would like to defect, but fears mutual defection. The other player has the deadlock preferences and prefers to defect no matter what (preferably with the other player cooperating). These two sets of preferences describe a game that looks like this:

		Deadlock Player: Cooperate	Deadlock Player: Defect
Chicken Player:	Cooperate	2, 1	**1, 3**
	Defect	3, 0	0, 2

One instance of bully is the biblical tale demonstrating the wisdom of King Solomon (1 Kings 3:16–28). Two women claim the same child as their own. One is the real mother, and the other is an impostor. Solomon proposes splitting the child in two. Hearing this horrific suggestion, one woman abandons her claim to the child. Solomon awards the child to that woman. The real mother would love her child so much she would give it up to save its life.

In other words, the true mother has the preferences of a chicken player. The knife is poised over the child. The dilemma is whether to stand firm (defect) or give in (cooperate). The real mother most wants to prevail—to stand firm on her claim while the impostor backs down. The worst outcome from the real mother's standpoint is for neither woman to give in. Then the child is cut in two.

The impostor has deadlock preferences. She evidently would rather see the child killed than have it go to her rival. The name "bully" tells what happens. The deadlock player can be a bully and defect. The chicken player is powerless to prevent this. The only thing she can do is cut her losses by cooperating. Ergo, the woman who gives in is the real mother.

Bully is a model for military confrontations in which one nation prefers to start a war, and the other views the war as a catastrophe to be avoided at all costs. To the extent this is an accurate model, the conclusion is gloomy. The belligerent nation will likely get its way, while the conciliatory nation is exploited to maintain the peace. The prognosis is probably worse yet. Real nations' preferences are fluid, and a nation that feels it has been exploited may decide that war isn't so bad after all.

JUSTIFYING COOPERATION

Morton Davis said that the average person's reaction to the prisoner's dilemma is not so much to ask what you should do but rather to ask how you justify cooperation. Going by the literature, that's been the reaction of a lot of game theorists, too. The literature of social dilemmas is rich in "solutions" and prescriptions, some of which exhibit the hopeful ingenuity of the theologian. The greatest number of articles discuss the prisoner's dilemma, but most attempts to justify cooperation apply to chicken and stag hunt as well. If there is any recurring theme in these arguments, it is that it is easier to evade social dilemmas than to resolve the paradox.

First there's the "guilt" argument. The temptation payoff of a prisoner's dilemma is tainted goods. It comes at the cost of betraying someone. You're better off cooperating—at least you'll sleep nights.

This analysis is fallacious for it introduces extra "payoffs" in the form of a clean or guilty conscience. Once again, the problem is confusion of utility and tangible gains. This issue is so important that it's worth thinking about carefully.

What if you were in a prisoner's dilemma experiment with prizes big enough to be meaningful to you and the other player—for instance, $5 million/$3 million/$1 million/0?

Were I in this situation, I might well cooperate, but I don't think the game would be a prisoner's dilemma at all. It wouldn't be because I'd prefer the mutual cooperation outcome to the unilateral defection outcome; $3 million is enough to buy virtually anything money *can* buy, so the extra $2 million confers little additional utility—less, probably, than the satisfaction I'd get from helping someone else win $3 million. The reason I'd consider defecting would be fear that the other person would defect. I'd still prefer $1 million to nothing at all, and wouldn't feel bad about defecting if I was sure the other person was going to defect. My decision would depend mainly on whether I thought the other person was going to defect, and thus would be more a matter of psychology than game theory.

This, however, is only half of a prisoner's dilemma. It has the element of fear but not of greed. Should the other player also prefer mutual cooperation to unilateral defection, the game would be a stag hunt. The stag hunt, however, is a much less troublesome game than the prisoner's dilemma. Rational players believing in each other's rationality cooperate in a stag hunt.

The point is that a certain set of tangible payoffs is not enough to guarantee that a prisoner's dilemma exists. For a person with a sufficiently strong sense of empathy, there is no such thing as a prisoner's dilemma. The dilemma arises when two persons' preferences are ordered a certain way. If your preferences never, ever, fit this pattern—if the "guilt" of betrayal outweighs the personal advantage gained *in every possible situation*—then you can never find yourself in a prisoner's dilemma. This is no more remarkable than saying that if you *never* prefer eating lunch, then for you there is no such thing as hunger.

Of course, this doesn't solve the riddle. The undeniable fact is that many people's preferences do create prisoner's dilemmas.

Some writers have minimized the prisoner's dilemma with the statement that communication is the "solution." The parties should communicate their intentions and come to an (enforceable) agreement to cooperate.

You won't get much argument that this is fine practical advice. It again takes us outside the purposely restricted scope of the prisoner's dilemma, though. To the extent that it is possible to confer beforehand and reach an enforceable agreement, the situation is not a prisoner's dilemma. The lack of communication, or more precisely, the lack of any way of enforcing a prior agreement, is the heart of the dilemma. A world of perfect communication and perfect honesty is a world without prisoner's dilemmas, but it is not the world we live in.

At the risk of being obvious: avoid prisoner's dilemmas whenever possible!

We have already touched on the most popular way of justifying cooperation, the "what if everyone did that?" argument. This may be broadened to suggest that your opponent actually will do, or is likely to do, whatever you do, so you'd better cooperate.

No one has worked harder at finding a cooperative solution than Anatol Rapoport. In *Fights, Games, and Debates* (1960), he says:

> *Each player presumably examines the whole payoff matrix. The first question he asks is "When are we both best off?" The answer in our case is unique: [mutual cooperation]. Next "What is necessary to come to this choice?" Answer: the assumption that whatever I do, the other will, made by both parties. The conclusion is, "I am one of the parties; therefore I will make this assumption."*

Many find this sort of argument appealing; others shrug. The counterargument might be put this way: when I was a child, my parents told me not to go into the basement because the bogeyman was down there. The stairs were rickety, and there was dangerous stuff down there, so maybe I was better off believing that there was a monster living in our basement. But just because you're better off believing something doesn't mean that it's true.

It would solve the problem if only everyone believed that the other players in a prisoner's dilemma would mirror their own choice. But not everyone does believe that, and the other player is under no compulsion to do what you do.

Others take this idea further. One clever argument holds that you should adopt the conscious policy of cooperating in prisoner's dilemmas because of—well, because of everything we know about prisoner's dilemmas. Forewarned is forearmed. Now that we recognize what a prisoner's dilemma is, and how people get in trouble following their individual rationality, we should resolve to cooperate every time we're in a prisoner's dilemma.

The idea is that the prisoner's dilemma falls into a gray area of logic. Neither cooperation nor defection is demonstrably right. Either you're going to be the kind of person who defects, or the kind who cooperates. Cooperators are better off (at least when in the company of other cooperators). That's why we should wise up and choose to cooperate as a matter of policy.

This line of reasoning too has achieved popularity and provoked skepticism. One question is whether a prisoner's dilemma can be called a dilemma for someone who has an iron-clad rule about cooperating in prisoner's dilemmas. If you *always* order liver and onions, you don't need a menu because there's nothing to decide. If you *always* cooperate, then you don't need to look at the payoff table—and perhaps it's wrong to say that you face a dilemma at all.

Recall how von Neumann was able to draw a parallel between matrices of numbers and human passions in the first place. Fundamentally, game theory is about abstract problems of maximization. Strictly speaking, the idea that games are conflicts between people is only a nice analogy. It is like an arithmetic primer that talks of adding two oranges to three oranges—really, arithmetic is not "about" oranges at all.

People want things, often numerical things like money or points in a game. Much of the time, people act to maximize those numerical things. Hence the analogy to numerical maximization. Where people don't faithfully maximize individual gain, the analogy fails. Game theory has nothing to say. A rational person who *consistently* foregoes the gains of defection when such actions can't influence the other player's choice is not maximizing his tangible winnings. There's nothing wrong with that; it just means that the payoffs in the table aren't the whole story. Something other than points is important.

The numbers in the table should reflect your net preferences, taking into account guilt, satisfaction derived from helping others, and even any sort of intellectual preference for cooperating in prisoner's dilemmas. If you prefer always to cooperate in prisoner's dilemmas, then that very preference distorts the dilemma out of shape. Arguably, a game in which one player prefers to cooperate "no matter what" is not a prisoner's dilemma.

HOWARD'S META-GAME

In 1966 Nigel Howard of the University of Pennsylvania offered a justification for cooperation that raised ingenuity to new heights. The difficulty with the prisoner's dilemma, according to Howard, was that the two strategies do not adequately express the range of intentions that players have.

What lies behind a defection? There is a world of difference between the "bad" defection of someone who ruthlessly defects as a matter of policy and the "nice" defection of someone who would really like to cooperate, and who *would* cooperate if only he thought the other player would. He's defecting to protect himself.

There's nothing much that can be done about a "bad" defector except to return his defection. Nor is the result of such defections so hard to accept. The tragedy is when two "nice" players defect because they misread the other's intentions. The puzzle of the prisoner's dilemma is how such good intentions pave the road to hell.

Howard's response was to create a "meta-game." The strategies of the meta-game are not cooperation and defection but rather *intentions* about how one would play the game. For instance, one "meta-strategy" is to "do what you think the other player will do." This is distinct from, say, "cooperate no matter what the other player will do." Howard produced a 4-meta-strategy by 16-meta-*meta*-strategy game matrix. He showed that mutual cooperation was an equilibrium point.

Howard's idea was so highly thought of by some that Anatol Rapoport could call it "*the* resolution" of the prisoner's dilemma paradox in 1967. It rated a *Scientific American* article heralding it as the end of the paradox. Enthusiasm has since faded. The meta-strategies are sleight of hand that don't really change anything. The "row player" and the "column player" in Howard's matrix are not the real, flesh-and-blood players but rather one player's guesses about how he and his partner are operating. Guess and second-guess all you want; ultimately, you either cooperate or defect, period. The last-instant temptation to defect remains.

None of these attempts have changed the consensus of game theorists. The rational thing to do in a one-time-only prisoner's dilemma *is* to defect. Defection is rational because it always results in a higher payoff (for a given choice by the other player) and because there is absolutely no possibility of one's choice having a causal effect on the other player's choice.

Yet people *do* cooperate, both in psychological experiments and in everyday life. Melvin Dresher thought that corporate price-fixing was a particularly clear-cut case of cooperation without communication. In a competitive industry (airlines are a good example) a company that charges a lower price attracts so much business it can make up for the lower profit in volume. Undercutting the competition is defecting, and maintaining the original high prices is cooperation (from the viewpoint of the companies, if not the public). Every so often, the airlines go through a cycle in which one company lowers prices, the others are forced to follow suit, a price war results, with all the airlines earning low or no profits, and then prices edge up again. Except during these price wars, airlines usually charge the same fares.

Executives aren't likely to call each other up and confer on what fare to charge. The Sherman and Clayton Antitrust Acts strongly forbid price-fixing among U.S. corporations. To avoid prosecution, companies go to such lengths as barring their executives from joining country clubs frequented by executives of their competitors. Dresher realized that this legislation creates a true prisoner's dilemma, and that companies cooperate (by maintaining prices that allow large profit margins) more often than not.

It's no exaggeration to say that society is founded on cooperation. Whether to litter—leave a tip—shoplift—stop and help someone—lie—conserve electricity—etc., etc.—all are dilemmas of individual gain and the common good. Some commentators have speculated that *irrational* cooperation is the cornerstone of society, and without it life would be, as Hobbes put it, "solitary, poor, nasty, brutish, and short." Game theorists might be pulling at the very thread of irrationality that holds the social fabric together. Without that thread, it all unravels.

Why do people cooperate?

There appears to be a very good reason why people and organizations cooperate. Most of the dilemmas of real life are iterated dilemmas.

Repeated interaction changes things greatly. Suppose you are a thief trying to sell a diamond and have the option of going through with the deal or cheating. It occurs to you that you might want to do business with the same buyer in the future. In that case, there is a much stronger argument for honor among thieves. If you cheat the buyer, that will nix any chance of consummating another transaction. You may well decide honesty is the best policy—purely out of self-interest.

BACKWARD INDUCTION PARADOX

In the Flood-Dresher experiment, Williams and Alchian cooperated for long stretches. And why not? They won reward payoffs every time they both cooperated. It's easy to see that that's the best anyone can expect to do *in the long run.* Players could do better in the present dilemma by defecting, but no rational opponent is going to let a player defect while he keeps cooperating. Defection is like a hot-fudge sundae. It tastes good now, but it's no good for you in the long run.

Look how the Flood-Dresher experiment ends (see pages 115–16). From Game 83 on, the players have learned to cooperate. Or have they?

In their comments, both players recognize the last game (Game 100) as a special case. In the long run—as long as there is a long run—both players ought to cooperate. But in the last game, why not grab for everything you can get? You don't have to worry about the other player retaliating in the future. There is no future.

In effect, the last game of an iterated prisoner's dilemma is a one-shot prisoner's dilemma. You ought to defect, just as you would in any other one-time-only dilemma.

Williams was inclined *not* to be greedy on the last game (comment to Game 10). You get the impression that was more a matter of chivalry than rationality. He would be at least a penny better off by defecting in the last round.

Alchian took this idea a step further. He expected Williams to defect in the last game, *and possibly earlier* (comment to Game 91). Alchian wanted to start defecting just before Williams would.

Hmm . . . If it's a given that rational players must defect in Game 100, then that means the next-to-last game is the last in which there is a meaningful choice between cooperation and defection. In other words, you might as well defect in Game 99. You don't have to worry about the other player getting mad and defecting in Game 100—we've already decided he's going to do that.

The ground crumbles beneath our feet! Evidently Game 98 is the last in which there's even a possibility of cooperating. But if that's so, you can defect in Game 98 without any qualms. Write off Game 98 and you might as well defect in Game 97. Ditto for Game 96—95—94 . . . all the way back to Game 1. You should defect on *every* dilemma. The upshot is that an iterated prisoner's dilemma isn't so different from a one-shot dilemma after all.

This is a distressing conclusion. It is so hard to accept, and so at odds with experience, that it has long been branded a paradox. This, the so-called "backward induction paradox," lies in the fact that "rational" parties get stuck with the punishment payoff every time, while less logical parties cooperate and do better.

Another thing that's hard to swallow is that the paradox is limited to iterated dilemmas of known, finite length. If the two parties don't know how many prisoner's dilemmas lie ahead of them, they can't apply the above reasoning. There is no "last" dilemma to reason backward from. In their ignorance, they have more grounds to cooperate. Also, if the number of dilemmas is infinite, there is no last dilemma and no paradox. Immortal beings can cooperate but not us mortals!

Game theorists' feelings about the backward induction paradox have been equivocal. The prevailing opinion has long been that it was "valid" in some abstract sense but not practical advice.

Here's one way to look at it. Von Neumann dealt exclusively with games whose rules were known in advance, as they would be in a parlor game. Faced with a new parlor game, a bright person might mull over the rules and decide how best to play. Von Neumann caricatured this into the operating assumption of game theory, namely, that players contemplate all possible courses of play and pick the best all-inclusive strategy *before the game's first move.*

It is more typical of the "games" of human interaction that we don't realize we are playing a game until we are in the middle of it. Even then, we may not know all the rules or how long the game will last. In a game of indefinite length, it's impossible to formulate all-inclusive strategies. The best you can hope for is a provisionary strategy, one mainly concerned with the next few moves. The actual "strategies" of chess players look forward a few moves rather than backward from a distant end game. You can say much the same thing about the strategies people use in their everyday lives. One reason that even self-serving people don't always betray others is that it's wise not to "burn your bridges." You can't say when you'll meet up with the same person again and need his cooperation.

Confusion about these two types of games and strategies is the heart of the backward induction paradox. Nash saw the Flood-Dresher experiment as a finite game of one hundred moves. Only the strategy of defecting a hundred times in a row guarantees that the player will not wish he had done something else, given what the other player did. Flood, on the other hand, described the experiment as a "learning experience." He didn't expect the players to treat it as a single extended game. He simply wondered if they would manage to cooperate after a certain amount of trial and error. Both views are valid, but Flood's better describes the subjects' actual mental processes.

In much of the later part of the Flood-Dresher experiment, Williams and Alchian adopted the policy of cooperating unless the other defected. As we will see in the next chapter, that's a very good way to play.

12

SURVIVAL OF THE FITTEST

In the 1980s game theory veered in a direction that even von Neumann may never have anticipated. Presently, some of the most actively pursued applications of game theory are in biology and sociology. Game theory offers appealing explanations for many types of biological cooperation and competition that previously were inexplicable.

Images of nature are often brutal ones. It's a jungle out there; it's dog eat dog and survival of the fittest. When one speaks of the lion laying down with lamb, one speaks of a supernatural event, not of ecology. Yet there is cooperation in nature.

Birds called ziczacs enter crocodiles' mouths to eat parasites. The crocodiles don't harm them. The crocodiles rid themselves of parasites and may learn of intruders from the birds' actions. In turn, the birds get a meal out of it. The crocodiles could easily "defect" by eating the birds. Why don't they?

Such instances of cooperation have long puzzled biologists. Sure, the arrangement is mutually beneficial. What if every crocodile ate ziczacs? Then there wouldn't be any left and there would be no birds to clean out the parasites. But that's an intellectual argument. Most biologists would be surprised to find that anything like that was going through a crocodile's head! It's equally hard to imagine that crocodiles have a moral code forbidding ziczac-eating. Why then does an individual crocodile (who may care nothing about "the good of species") pass up an easy meal?

STABLE STRATEGIES

The catch phrase "survival of the fittest" is prone to misinterpretation. It sounds like nature has a conscious program of selecting those species that are strongest, smartest, most prolific, or most vicious.

It's easy to fall into this trap because humans happen to be one of the rare species that *are* at a conspicuous pinnacle of ability. We are the most intelligent and technologically inclined species on the planet. This encourages the flattering notion that we are the unique goal of evolution. Yet every species that is around today has a lineage just as ancient as the human race's. Most other species are not conspicuously smart, strong, prolific, vicious, or *anything*.

In the very broadest sense, the doctrine of natural selection means that the observed phenomena of nature tend to be *stable* phenomena. We see stars in the sky because stars last a long time. Stars aren't smart and they can't reproduce, but they persist for billions of years in a form stable enough that we recognize them as a class of natural phenomena.

On earth, few physical structures survive indefinitely. Nonetheless, the processes of nature destroy mountains, rivers, clouds, and ocean waves at very nearly the same rate that they produce new ones. The new mountains, rivers, clouds, and waves are so similar to the old that we recognize them as being of the same kind. In this way, many non-living phenomena persist.

Another way for a class of phenomena to persist is to produce copies of themselves. This is what living things do. Individuals die, but others of their kind remain. The organisms we see today are those whose genes have been most successful in producing and preserving copies of themselves down through the generations.

The genetic code allows organisms to pass on not only physical traits but behaviors. Those genetically encoded behaviors that cause individuals possessing them to survive longer or produce more offspring are likely to persist.

The most familiar type of cooperation in nature is parental behavior. It at least is easy to account for. Birds with a hereditary instinct to build nests and feed their chicks are likely to produce more surviving offspring than birds without such instincts. In contrast, a bird that "defects" by ignoring its offspring might thrive and live to a ripe old age, but the genetic basis of this defection would be less likely to be passed on to succeeding generations. In the long run, the genes for nurturing would supplant the genes for shirking.

But this a special case. The birds are helping the very offspring that will carry their genes into future generations. Cooperation between unrelated individuals is something else again. In many ways, defection pays.

Suppose there was a species that shares by instinct. There is a certain fixed food source for the species, and they manage to dole it out evenly among them, saving a bit for the future in case of drought. This is a collectively rational behavior that provides for the needs of the group.

In nature, mutations occasionally change the genetic code. Over many generations, occasional individuals would be born without the gene for the sharing instinct. Call these individuals "gorgers." Gorgers always eat their fill. They don't share and make no effort to save food for the future. The behavior of the gorgers leaves less food for the sharers. (Any leftover food is communal property and is not assigned to those who ate less.)

Every few years, drought diminishes the supply of food. Bad droughts cause many of the species to starve to death. The gorgers hardly ever starve since they grab all the food they can. The drought kills come almost exclusively from the ranks of the sharers. So after every drought, the percentage of gorgers in the population is a little higher. The survivors give birth to a new generation that is also higher in the gene for gorging.

After many, many generations, you would expect that virtually all members of the species would be gorgers. The sharers might become extinct.

You can say the gorgers are "fitter" than the sharers. They are in the narrow sense that they eventually prevail. But "fit" is a misleading term. The eventual population of gorgers is not any better at surviving drought than the original population of sharers was. It's *worse.* Sharing and saving food for the future is probably the best way to survive a bad year. The thing is, gorgers do better than sharers *when both coexist.* Once the sharers become extinct, though, the gorgers lose their advantage.

Is it possible that mutations would then produce a few sharers and the population would return to sharing? No. There's no advantage in being a lone sharer in a population of gorgers. Any sharers born in the population of gorgers would die in the first bad drought.

Here gorging is an approximate example of what biologists call an "evolutionarily stable strategy." This is a genetically transmitted behavior that persists when nearly all members of the population behave this way because no alternate behavior (produced by a mutation) can displace it.

Sharing (in the context above) is *not* evolutionarily stable because a few gorgers can "take over" a population of sharers. The behaviors we expect to see in the world are evolutionarily stable strategies.

Evolutionarily stable strategies are not necessarily "rational" or "fair" or "morally right." They are simply *stable.*

IS DEFECTION IN THE GENES?

There are prisoner's dilemmas (and especially, free-rider dilemmas) in nature. They occur whenever an individual's self-interest is opposed to group interest. Members of the same species share the same basic needs. When food, water, shelter, or mates are at a premium, it is to be expected that an individual's gain may be a group's loss.

The hypothetical example of animals sharing a fixed food supply is likely to be a free-rider dilemma. Let's make this explicit with a game-theoretic table:

	Most of the other animals eat their fair share	**Most of the other animals gorge**
Eat fair share	(2, 2) All get some food, and there is a little left over for the future	(0, 3) I get less food than the others, and there's nothing left for the future; the others get a little extra because I didn't eat much
Gorge	(3, 0) I get plenty of food, and there is a little left over; the others are cheated slightly by my gorging	(1, 1) All get plenty of food, but there's nothing left for the future

The numbers in parentheses show the preferences of any particular animal and of the rest of group. The dilemma requires only two simple assumptions: that, all things being equal, each animal would ideally prefer *not* to gorge itself but rather to eat a little now and save some food for future needs (it would do that if it had its own private cache of food); and that, overriding the former preference, each animal *much* prefers whatever course will give it the more food.

Then the best outcome for any individual is to be the lone gorger: let the others show restraint and leave some food for the future! The worst is to be the only nongorger; then the others eat your extra food.

You might question this talk of "preferences" in dumb animals. How do we know what they prefer?

Game theory need not deal in preferences at all. Suppose we say that the numbers in the above table represent "survival value." When starvation threatens, an animal is most likely to survive if it gorges itself now and still has some food left over, thanks to the restraint of its fellows (lower left cell). It is next most likely to survive when all the animals show restraint and save food. Odds of survival are less when no food is saved, and are worst when the animal doesn't even get its fill now.

Natural selection "chooses" or "prefers" the behaviors that will maximize survival value. This is all we need to apply the mathematics of game theory, even though no conscious choices or preferences may be involved. Those animals that get the highest "scores" are most likely to survive and reproduce. The gorgers will survive at the expense of sharers and displace them. Once again we have a prisoner's dilemma in all its rational irrationality. The sensible strategy of sharing loses out to gorging, and all are worse off.

Evolution might be expected to instill many other types of defection behavior. Humans, of course, have been shaped by evolution as has every other species. Here then seems to be an explanation for the whole catalogue of human folly. Defection is an evolutionarily stable strategy. Cooperation is not. That's the way the world is, and the way people are. As von Neumann once asserted, "It is just as foolish to complain that people are selfish and treacherous as it is to complain that the magnetic field does not increase unless the electric field has a curl."[1]

Is defection "in the genes"?

The question is a complex one. To be sure, it does not follow that any type of defection has a biological basis. Human preferences and survival value are often quite different things. An awful lot of human preference centers on money, which means little to survival or fertility rates. Even so, greed for money and other material things might be a

1. The latter part of this aphorism is a law of physics familiar enough to von Neumann's listeners. In other words, "It is just as foolish to complain that people are selfish and treacherous as it is to complain that the earth is round."

side effect of genes that encourage self-interest in matters of food, water, and mates.

There can be more than one evolutionarily stable strategy. There are only the two strategies in a one-shot prisoner's dilemma. But an iterated prisoner's dilemma allows any number of strategies. A strategy for an iterated dilemma tells what to do (cooperate or defect) in each dilemma of a series, based on what the other player(s) did in previous dilemmas. Strategies of great subtlety are possible.

Most of the dilemmas in nature *are* iterated. Animals sharing a food source will face the same gorge-or-share dilemma, with the same group, many times. Thus the iterated prisoner's dilemma is of principal interest to biologists.

ROBERT AXELROD

By far the best-known study of iterated prisoner's dilemma strategies is a group of computer "tournaments" conducted in 1980 by Robert Axelrod, professor of political science at the University of Michigan. Reported in the *Journal of Conflict Resolution* and later in Axelrod's book *The Evolution of Cooperation* (1984) this work ranks among the most significant discoveries of game theory.

Axelrod came to game theory by a circuitous route typical of the field. He was a math major at the University of Chicago, where he took Morton Kaplan's class. On a recommendation from a friend, he read R. Duncan Luce and Howard Raiffa's *Games and Decisions.* The first sentence of the book hooked him: "In all of man's written record there has been a preoccupation with conflict of interest; possibly only the topics of God, love, and inner struggle have received comparable attention."

Axelrod went to Yale to get a doctorate in political science. His thesis was on conflict of interest, a topic Luce and Raiffa never quite defined. Today Axelrod is associated with the University of Michigan's Institute of Public Policy Studies. His title of Professor of Political Science and Public Policy belies professional interests that range from biology to economics.

What makes the iterated prisoner's dilemma different from a one-shot dilemma is the "shadow of the future," to use Axelrod's colorful phrase. It makes sense to cooperate now in order to secure cooperation in the future. But no one takes the future quite as seriously as the present. Players subjectively weight a present advantage against possible future losses. "A bird in the hand is worth two in the bush."

When only the present dilemma is subjectively important, it is effectively a one-shot dilemma in which defection is to be expected. But when players value future as well as present gains, a true iterated dilemma exists. Then a number of conditional strategies are possible.

Axelrod invited a number of well-known game theorists, psychologists, sociologists, political scientists, and economists to submit iterated prisoner's dilemma strategies for a tournament to be conducted by computer. The computer, of course, was just for the sake of expediency. The tournament might as well have been conducted as a game with people sitting around a table and following their prescribed strategies to win poker chips.

Each strategy specified what to do on the present dilemma of a series, given the complete history of the interaction—what the strategy and its opponent had done up to then. For each dilemma, the payoff was in points:

	Cooperate	Defect
Cooperate	3, 3	0, 5
Defect	5, 0	1, 1

Axelrod's tournament was organized like a round-robin tournament for tennis or bowling. Each program was matched against all the other programs submitted, against itself, and against a program that chooses cooperation or defection randomly. This thorough mixing was necessary because strategies have different "personalities." For instance, the strategy of always cooperating does well when matched with itself, but it gets the worst possible score when paired with the strategy of always defecting. Axelrod tallied how well each strategy did against all the other strategies.

Each iterated dilemma consisted of 200 separate dilemmas. Since it was possible to win as much as 5 points on each, the scores could theoretically range from 0 to 1,000. These extremes would obtain if the strategy of always cooperating was paired with always defecting. Then the former would get 0 points every time, and the latter 5 points. Neither of these simplistic strategies was submitted, however.

A realistic good score would be 3 points for each of the 200 rounds

(600 total). Two strategies that managed to cooperate every time would get that. Conversely, 200 points is pretty bad because that's what a strategy can guarantee itself just by playing safe and always defecting. The trick is to gain extra points by winning the reward or temptation payoffs part of the time. Each strategy's score varied depending on the strategy paired with it. To get an overall score, Axelrod averaged each strategy's scores.

Let's examine a few ways of playing an iterated prisoner's dilemma. Axelrod sometimes does this in class by having two students play each other. Standing some distance apart at the blackboard, the students choose simultaneously by writing a "C" for cooperate or a backward "C" (which sounds the same written on a blackboard) for defection. The demonstration quickly points up some of the hazards of common strategies. Sharp players quick to defect may get an initial windfall but are apt to get stuck in a rut of punishment payoffs thereafter. Nice players who cooperate often get stung with a sucker payoff and find it difficult to convince their partner to reciprocate.

One of the simplest strategies is:

- *Always defect* ("ALL D"). You never give a sucker an even break by defecting on each and every round no matter what. This is just what the backward induction paradox counsels. It's the safest strategy, in that no one can possibly take advantage of you.

The opposite strategy is:

- *Always cooperate* ("ALL C"). If everyone plays this way, everyone comes out pretty well. Cooperating players get the reward payoff every time, and that's as good as anyone can expect in the long run. The catch is, the other person may not be so nice. Then an ALL C player is forever turning the other cheek—and getting slapped.

Another possibility is:

- *Cooperate or defect at random* ("RANDOM"). There's not much to be said for this, but let's toss it out as another possibility. The RANDOM program Axelrod included cooperated 50 percent of the time.

None of these strategies are nearly as good as they could be. All are "blind" strategies, fully prescribed in advance. They do not take into account what the other player has done. The advantage of doing so is overwhelming.

Suppose that after dozens of rounds, you become convinced that the other person is playing an ALL C strategy. It is as if the other player is on autopilot, and can be counted on to cooperate no matter what. If so, it's your lucky day. Faced with an opponent who always cooperates, you should defect. You can expect to win the temptation payoff every time, and this is the best possible outcome of an iterated prisoner's dilemma.

Suppose instead that the other person is always defecting, no matter what you do. Then you should always defect. That at least gets you the punishment payoff, which is better than the sucker payoff.

These two cases are the simplest of all the possibilities. In general the other player will react to *your* actions, too. The question is, what is the best way for you to play knowing that it will affect how the other person plays?

This question probes the possibilities of communication by actions. The players are not allowed to send messages to each other. They cannot pass notes, make deals, or sign treaties. On the face of it, this is a considerable limitation. It would be helpful to have a caucus, as in one of the Ohio State studies; to be able to say: "Let's be reasonable. The reward payoff is the best that either of us can expect, so let's cooperate, even on the last dilemma." Or: "Cooperate or else! If you defect even once, I'll defect from then on."

The players can "speak" only with their actions. This restricted language permits some types of deals or threats to be communicated more readily than others.

The cooperate-or-else threat above is tough to make through actions alone. The fact that a threat has been made isn't apparent until the other player defects. Then you have to defect for the rest of the iterated dilemma to make good on the threat. There is no provision for the other player to mend his ways. Once he defects, he has no incentive to do anything but defect. Far from enforcing cooperation, a cooperate-or-else strategy is likely to lead to constant mutual defection.

TIT FOR TAT

Fourteen strategies were submitted for Axelrod's first tournament. Some were fairly complex programs. The longest had seventy-seven lines of computer code. That program also did the worst of any submitted by a correspondent, an average of 282.2 points. (The RANDOM strategy did slightly worse with 276.3 points.)

The highest score went to the simplest strategy. Submitted by Anatol Rapoport, it was called TIT FOR TAT. TIT FOR TAT was a strategy known to work well in tests with human subjects.

TIT FOR TAT took just four lines of computer code. It can be explained in a sentence of plain English: cooperate on the first round, then do whatever the other player did on the previous round.

Why is TIT FOR TAT so effective? For one thing, TIT FOR TAT is a "nice" strategy. In the jargon of game theory, a nice strategy is one that is never the first to defect. TIT FOR TAT starts out by cooperating: it gives its opponent the benefit of the doubt. Should the other strategy return the favor and continue to do so, TIT FOR TAT never defects. TIT FOR TAT never makes trouble and is content to leave well enough alone. In particular, when TIT FOR TAT is paired with itself, both sides start out cooperating and never have any provocation to do otherwise.

But a strategy that is too eager to cooperate often gets clobbered. TIT FOR TAT is also provocable. It defects in response to defection by the other strategy. Remember, after the first round, TIT FOR TAT echoes whatever the other strategy did. If the other strategy defects on Round 5, TIT FOR TAT defects on Round 6. That gives the other strategy an incentive to cooperate.

Just as important is TIT FOR TAT's "forgiveness." The other strategy can't "learn" that it has an incentive to cooperate until it defects and gets punished for it. TIT FOR TAT isn't so draconian that a single transgression leaves its opponent with no incentive to cooperate ever again. TIT FOR TAT is "willing" to start cooperating any time its partner strategy is.

That the winning strategy was nice and provocable was no surprise to the game theory community. Many of the strategies submitted were both nice and provocable. A third attribute was a surprise. TIT FOR TAT is *simple.*

TIT FOR TAT threatens, "Do unto others as you would have them do unto you—or else!" No strategy is allowed to deliver a threat in so many words. The threat has to be implicit in the strategy's behavior. TIT FOR TAT does this by repeating the most recent action of its opponent. This is in the "hope" that the other strategy will "realize" what TIT FOR TAT is doing. If it does, then it will conclude that it is only hurting itself by defecting.

Not all strategies are capable of reacting to TIT FOR TAT's threat. However, by making its threat as simple as possible, TIT FOR TAT ensures that the maximum number of responsive strategies will "understand" it. A rule of dog training is that you should punish a dog immediately after it has done something wrong. Like dogs, strategies may have short attention spans. TIT FOR TAT "punishes" for a defection immediately, on the very next dilemma. This gives it an advantage over a more sophisticated strategy that, so to speak, counts to ten before it gets angry (such as by letting a certain number of defections go by before retaliating). To be sure, a very clever strategy might understand that another strategy was doing this. Not all strategies are clever, though. TIT FOR TAT succeeds because its threat is the simplest possible and thus the easiest to react to.

Still another important quality of TIT FOR TAT is that it does not have to be kept a secret. Someone playing TIT FOR TAT need not fear that his opponent will surmise that he is playing it. Far from it—he *hopes* that the opponent will realize it. When faced with TIT FOR TAT, one can do no better than to cooperate. This makes TIT FOR TAT a very stable strategy.

TIT FOR TAT scored an average of 504.5 points. Against specific strategies, its scores ranged from a low of 225 points to a high of 600 points.

Axelrod's first tournament was not conclusive proof of the merit of TIT FOR TAT. A strategy's score depends greatly on the other strategies, and the assortment of strategies in the tournament, while diverse, was not necessarily representative of the universe of possible strategies. They were strategies chosen by professionals to get a high score, reflecting then-current views about how to play an iterated prisoner's dilemma. Those views may or may not be correct, nor need they be "typical" of all possible strategies. It was conceivable that TIT FOR TAT could simply be a strategy that does well against how top theorists think the game *ought* to be played.

Axelrod held a second tournament in which the entrants were informed of the results of the first tournament and knew how well TIT FOR TAT did. The implicit challenge was to beat TIT FOR TAT. In all, he got sixty-two entries from six countries. Despite the fact that many were trying to beat it, TIT FOR TAT was again the winner. This was strong evidence that TIT FOR TAT was optimal or nearly so when matched with a wide variety of opponent strategies.

To Axelrod, one of the more surprising findings was that TIT FOR TAT won without ever exploiting another strategy. "We tend to compare our scores to other people's scores," he explained. "But that's not the way to get a good score. TIT FOR TAT can't beat anybody, but it still wins the tournament. That's a very bizarre idea. You can't win a chess tournament by never beating anybody."

TIT FOR TAT is a good strategy with other social dilemmas, too. An iterated stag hunt works much like an iterated prisoner's dilemma. Repetition boosts the incentive to cooperate. Surprisingly, the game of chicken becomes *more* troublesome when iterated.

Why? You're playing chicken, only now you have a reputation to consider. Say you've decided that the way to play chicken is to swerve (a policy most people would agree with). You can't get the best payoff that way, but it's the only way to make sure that you don't crash and die. Given the situation, you've resigned yourself to swerving as the only "reasonable" course.

The trouble is, this gives you a reputation as a "chicken" who always swerves. Then your opponent can feel safe driving straight down the middle of the highway. He gets the best-possible payoff while you get the next-to-worst. *Something* is wrong, since your positions at the beginning of the iterated game were identical.

It works the other way, too. If you can build a reputation as a tough guy who never swerves, then everyone else will figure it's suicide not to swerve. Iteration therefore seems to promote the strategy of *not* swerving. But if everyone (conscious of their reputation) resolves not to swerve, everyone crashes.

Only a conditional strategy like TIT FOR TAT allows the iterated chicken player to put up a reasonable front: to be a nice guy and swerve but still offer an incentive for the other player to swerve, too. The threat of defection is a powerful one in chicken. If the other player defects, you lose. Everyone has a vested interest in convincing his partner not to defect. When the partner is playing TIT FOR TAT, the only way to ensure cooperation is to cooperate in kind. The present gain from defection is canceled by the loss from the other player's future defection.

THE TROUBLE WITH TIT FOR TAT

As well as TIT FOR TAT performed, it does not follow that it is the "best" of all possible strategies. It is important to realize that no strategy is good or bad out of context. How well a strategy does depends on the strategies with which it interacts.

TIT FOR TAT does have several failings. It doesn't take advantage of unresponsive strategies. When paired with ALL C, TIT FOR TAT cooperates and wins 3 points each dilemma. It would do better to defect and win 5 points. In fact, with any unresponsive strategy, the best course of action is to defect. Defection always yields a higher payoff in the current dilemma, and there is no possibility of retaliation with an unresponsive strategy.

TIT FOR TAT is more or less predicated on the assumption that the other player is trying to get a good score. After the first move, TIT FOR TAT repeats the strategy of the other player. When paired with a "mindless" strategy like RANDOM, TIT FOR TAT descends to its level and does no better.

Still another problem with TIT FOR TAT is that it is subject to "echo" effects. Among the strategies submitted to Axelrod were some that were very similar to TIT FOR TAT. They differed in that they tried to better it by defecting occasionally. Suppose that one of these strategies (ALMOST TIT FOR TAT) is paired with true TIT FOR TAT. Both are cooperating. Then out of the blue, ALMOST TIT FOR TAT throws in an uncalled-for defection. This causes TIT FOR TAT to defect on the next dilemma. Meanwhile, ALMOST TIT FOR TAT has cooperated as usual. On the dilemma after that, ALMOST TIT FOR TAT will echo the previous defection. The two strategies will alternate defecting and cooperating indefinitely. The pattern looks like this:

TIT FOR TAT: C C C C C D C D C D C D . . .

ALMOST TIT FOR TAT: C C C C D C D C D C D C . . .

In an echo, each strategy averages 2.5 points each dilemma (the average of the sucker payoff, 0, and the temptation payoff, 5). This is less than the 3 points each strategy wins during mutual cooperation.

Echo effects are not a glitch of one particular, rigid strategy. How many real conflicts find both sides claiming that they are only retaliating for attacks on themselves?

Some of these problems are easier to fix than others. It appears to be possible to improve on TIT FOR TAT's results with unresponsive strategies. Pretend you're playing an iterated prisoner's dilemma against an unknown opponent (as in a psychological experiment). We already know that TIT FOR TAT is a good strategy, so let's say that you start by following TIT FOR TAT to the letter. Only after you've had a chance to size up the opponent's play do you consider deviating from TIT FOR TAT. Even then, you deviate from TIT FOR TAT only if you think you can do better.

When dealing with an ALL C player, it makes sense to defect and get the temptation payoff rather than just the reward. The trick is distinguishing an ALL C player from another player using TIT FOR TAT or a similar strategy. Playing TIT FOR TAT with itself or ALL C results in an unbroken string of cooperations. The only way to find out if a cooperating strategy is ALL C is to try something different—namely, defect.

You might try an ALMOST TIT FOR TAT strategy that, after a hundred rounds with a perfectly cooperative partner, defects just once. Then you watch to see what the partner does on the round after that. (You cooperate on that round.) If the other strategy lets you get away with this by not defecting, then you start defecting. You defect ever after, or at least until the other strategy defects. If the other strategy defects in response to your defection (as TIT FOR TAT would) then you go back to cooperating and are only out a point for trying. (You win 5 points when you defect and nothing on the next round when your partner "punishes" you. Otherwise, you would have won 3 points on both rounds for cooperating.) Notice that you avoid the echo effect mentioned above by not retaliating for TIT FOR TAT's "provoked" defection.

This scheme would let you take advantage of ALL C. It's almost as good as plain TIT FOR TAT when played against TIT FOR TAT, as you are penalized only 1 point for testing the other strategy. If you had reason to believe that the other strategy was going to be either ALL C or TIT FOR TAT, with approximately equal probability, then this would be a good way to play.

But is it reasonable to think that the other strategy is likely to be ALL C? The working premise of game theory is that the other player will choose the best possible strategy open to him. From what we already know, ALL C isn't a very good strategy at all. It is wishful thinking to believe that your opponent will use ALL C, or even to think there's a meaningful chance he will use it. Therefore, we shouldn't place too much stock in the fact that a particular strategy does well when matched with ALL C.

Regular TIT FOR TAT is a more likely opponent strategy. We know it did well in Axelrod's tournaments. However, the ALMOST TIT FOR TAT strategy, which throws in a test defection to see if it's dealing with ALL C, is *not* as good as plain TIT FOR TAT when paired with TIT FOR TAT. It's 1 point worse. Beating TIT FOR TAT is tougher than it looks.

Axelrod's tournaments included sophisticated strategies designed to detect an exploitable opponent. Some created a constantly updated statistical model of their opponent's behavior. This allowed them to predict what the opponent strategy will do after cooperation and after defection, and to adjust their own choices accordingly.

This sounds great. It does allow these strategies to exploit unresponsive strategies like ALL C and RANDOM. The trouble is, no one submitted a unresponsive strategy (other than the RANDOM strategy Axelrod included). No such strategy did better than TIT FOR TAT overall. These "sophisticated" strategies are like Swiss Army knives that have so many different tools you never need that they are too heavy to carry. Most successful strategies aren't very exploitable, and the act of provoking them to see how they react usually costs more points than are to be gained.

One key to TIT FOR TAT's success is that its punishment for defection is commensurate with the crime: a defection for a defection. That may not be the optimum arrangement. Would it work better with harsher or more lenient punishment?

As an example of a stricter version, you could play "TWO TITS FOR TAT" and defect twice following any defection. A more lenient version would be "TIT FOR TWO TATS" that would ignore isolated defections and punish only after two consecutive defections.

TIT FOR TWO TATS, the more forgiving strategy, does respectably well. It was not entered in Axelrod's first tournament, but he reported that, had it been, it would have won.

TIT FOR TWO TATS was submitted in the second tournament. There it did only about average, and far worse than TIT FOR TAT. This suggests that when strategies are designed to do well with TIT FOR TAT, TIT FOR TWO TATS is rather too lenient.

TWO TITS FOR TAT errs in the other direction. When paired with a near-TIT FOR TAT strategy that occasionally defects, the defection sets off a worse-than-usual echo effect. Both strategies end up defecting ever after.

Echo effects can be minimized by a strategy like "90 PERCENT TIT FOR TAT." This strategy is like TIT FOR TAT except that it retaliates for a defection with a probability of 90 percent. Occasionally it lets a defection go by unpunished.

This defuses echo effects. Defections volley back and forth like the ball in a Ping-Pong game until one strategy misses a defection because of the 90 percent rule. Then both return to mutual cooperation. A strategy like 90 PERCENT TIT FOR TAT is useful in an uncertain environment—one where information about the other player's moves is garbled. The more reliable the information is, the higher the probability of retaliation should be set.

ARTIFICIAL SELECTION

The most thought-provoking of Axelrod's computer experiments was the third. Axelrod wondered what would happen if his computerized strategies were subjected to a kind of artificial "natural selection."

An iterated prisoner's dilemma strategy is a pattern of behavior, a personality. Think of a colony of living cells in a primordial soup, each with its own genetically encoded behavior. Every now and then two nearby cells interact. The interaction forces a choice between individual gain and the common good. The cells are sophisticated enough to "remember" each other and how they acted on previous interactions. Let the points won represent survival value. How well a cell does depends both on its behavior and on its environment—that is, on how well its strategy does when matched with the strategies of the other cells around it.

Axelrod ran a series of tournaments (an *iterated* iterated prisoner's dilemma) starting with an assortment of strategies. After each tournament (itself consisting of many rounds of the prisoner's dilemma), strategies would "reproduce" in computer simulation, the number of offspring being tied to the points each strategy scored. Thus the next tournament would start with a new generation including more copies of the successful strategies and less of the unsuccessful ones. Over many such generations, Axelrod hoped to see the more successful strategies become more common while less successful strategies would die out. The ultimate winners would be those strategies that did best against other successful strategies and against themselves.

During the first few generations of this simulation, weak strategies like RANDOM quickly died out. Meanwhile, other strategies became more common. These included not only TIT FOR TAT and similar strategies but also some highly exploitative strategies.

Then an interesting thing happened. After several generations, these "predator" strategies ran into trouble as their "prey" (exploitable strategies) became extinct. They began to "starve." When paired with TIT FOR TAT or other unexploitable strategies, predator strategies fell into cycles of mutual defection and scored poorly. The predators' numbers dwindled. Eventually, TIT FOR TAT became the most common strategy in the simulation.

This is the strongest evidence yet that TIT FOR TAT is a naturally superior strategy—very close to the biologist's idea of an evolutionarily stable strategy.[2] TIT FOR TAT not only does very well when paired with itself, but it exerts pressure on other strategies to cooperate. In a population made up almost entirely of TIT FOR TAT players, one cannot do better than to play TIT FOR TAT oneself. Any defection, against any player, anytime, will be punished, so the most profitable course is to cooperate all the time (unless you meet someone who defects first).

Axelrod has theorized convincingly how cooperation might arise in unfavorable environments. TIT FOR TAT does poorly against highly exploitative strategies like ALL D. But if a "colony" of TIT FOR TAT

2. Technically, TIT FOR TAT isn't quite an evolutionarily stable strategy. By definition, such a strategy is stable against mutations. Because it always cooperates with nice strategies, TIT FOR TAT can't distinguish itself from ALL C and other nice strategies. If such strategies arose by mutation in a population of TIT FOR TAT, they could coexist indefinitely as they would win as many survival points. Conceivably, some strategies might even be slightly more successful than TIT FOR TAT and could eventually take over. What seems established is that there is an evolutionarily stable strategy for the iterated prisoner's dilemma that is very TIT FOR TAT-like. Most of the time, it would do what TIT FOR TAT does.

players can manage to have most of their interactions with other members of the colony, they can do better than the ALL D players.

Suppose a population is composed almost entirely of ALL D players, with just a few TIT FOR TAT's as well. At the outset, the ALL D's exploit the TIT FOR TAT's. After a while, almost every TIT FOR TAT has interacted with every ALL D at least once. In subsequent interactions, the TIT FOR TAT's echo the ALL D's defection and essentially act just as they do.

Normally, then, the ALL D players score 1 point per dilemma. They score this even when interacting with TIT FOR TAT players, for after the first interaction between strategies, the TIT FOR TAT player defects too.

The TIT FOR TAT players likewise score 1 point every time they interact with the ALL D players (after getting stung on the first interaction). But every time a TIT FOR TAT player interacts with another of its kind, they cooperate, and both win 3 points. This means the TIT FOR TAT players average a little better than 1 point per interaction. The numbers of TIT FOR TAT players will increase relative to the ALL D's. The more TIT FOR TAT's in the population, the greater the advantage in playing TIT FOR TAT. In this way, a small colony of TIT FOR TAT players could supplant a population of uncooperative players.

THE FISH IN THE MIRROR

Many species show cooperative behavior in the face of individual incentives to defect. What preserves such cooperation? Axelrod's findings are only suggestive, but they show that *conditionally* cooperative (TIT FOR TAT-like) behavior can hold its own against simple defection. Biologists have since attempted to demonstrate TIT FOR TAT-like behavior in cooperative species.

One of the simplest and most elegant experiments was conducted by German biologist Manfred Milinski. Milinski was intrigued by the cooperative behavior of the small fish known as sticklebacks. When threatened by a larger fish, sticklebacks send out a "scouting party" to approach the bigger fish closely and then return. Biologists believe such "predator inspection visits" help the sticklebacks identify the larger fish and tell how hungry or aggressive it is.

These predator inspection visits are like a dare game. A party of two or more sticklebacks approaches a predator fish closer than a single stickleback would. A predator can attack only one fish at once, so there is safety in numbers. Provided the sticklebacks stay together, they split the risk while gaining as much or more information.

How could this behavior evolve by natural selection? Go back to a time before the ancestors of the present-day sticklebacks had this behavior. Mutation creates a "curiosity" gene that causes a fish to approach predators on its own. A fish is born with this gene, it sees a bigger fish, and it goes to check it out. Then—CHOMP!—that fish is out of the gene pool.

The behavior makes sense only when there is cooperation between sticklebacks. Even then, there is always a temptation to be a free rider. A stickleback who hangs back and lets the others approach the predator minimizes its own risk while still getting the benefit. Natural selection would seemingly favor the "cowards" over the "leaders," and eventually all the fish would be such cowards that the behavior would vanish altogether. Not only it hard to see how predator inspection visits evolved; it is difficult to see how the behavior is stable once evolved.

Milinski noted that sticklebacks approach potential predators in short, cautious stages. They swim forward a few centimeters, then wait a moment. Normally, the other members of the scouting party follow suit. Each stage of the inspection visit is like a prisoner's dilemma. At the risk of putting thoughts into the head of a stickleback, Milinski guessed that each stickleback most "prefers" to hang back and let the others approach the predator. That way it avoids the risk of attack while gaining the information the others learn. Second choice for the stickleback is to cooperate with the others and go along with the predator inspection, shouldering its fair share of the risk. This is evidently preferred to the case where all the sticklebacks defect and call off the predator visit. Otherwise, the sticklebacks wouldn't attempt predator visits. The least preferred outcome is that where a stickleback's companions desert it and it faces the risk alone.

Because the inspection visit takes place in stages, it is an iterated dilemma. At each step of the way, the sticklebacks can reassure themselves (again, excuse the anthropomorphism) that their fellows have not deserted them. The sticklebacks approach the predator closely only when all have cooperated repeatedly.

Milinski's clever experiment used a mirror to simulate both cooperating and defecting sticklebacks. A single stickleback was placed at one end of a rectangular tank. At the other end of the tank was a glass panel behind which a large fish (a cichlid) was visible. This fish resembles the perch, whose diet consists largely of sticklebacks.

For part of the experiment, a mirror was placed along one long side of the rectangular tank. Sticklebacks being innocent of such things as mirrors, Milinski expected that the fish would perceive its mirror image as another stickleback. The fish in the mirror was a perfect reciprocator, of course. Every time the fish moved toward the potential predator, the mirror image moved an equal distance.

Another mirror simulated defection. For these trials, the "cooperation" mirror was removed and replaced with a mirror on an angle. This "defection" mirror was positioned so that the fish's image appeared close when the fish was at the end of the tank far from the predator. As the fish approached the predator, however, it would see its "cowardly" image turning tail and running off at nearly right angles.

Milinski found that the behavior of the sticklebacks was consistent with TIT FOR TAT. The fish started by cooperating—by darting a few centimeters toward the suspected predator—and then tended to do whatever their mirror image appeared to be doing. With the cooperating mirror in place, the image followed the stickleback, and this emboldened the fish to approach the larger fish closely. With the defection mirror, the sticklebacks halted their approach or continued very cautiously, as they sometimes did when alone.

Another animal that has acquired modest fame in scientific circles for TIT FOR TAT-like behavior is the vampire bat of tropical America. The vampire bat's table manners do not recommend it to human sensibilities. Gorged bats sometimes regurgitate blood into the mouths of roost mates who have not found a blood meal. This altruistic behavior, sometimes observed among bats that were known not to be closely related, has long challenged biologists. Like airliners with a full fuel tank, gorged bats, heavy with blood, consume more calories keeping aloft. They need the extra food less than the hungry bats, which, with their swift metabolism, quickly starve. A bat that fails to eat two nights in a row starves to death. But what stops bats from accepting surplus food without giving it? "Deadbeat" bats ought to have a higher survival rate and supplant the altruists.

Gerald S. Wilkinson, then of the University of California at San Diego, did a study in which bats were caged before feeding time and returned hungry to the roost. He showed that a bat that had been fed by another bat was more likely to donate blood subsequently. The willingness to donate depended primarily on how well known the bats were to each other. They were more likely to cooperate with familiar bats that they could expect to interact with in the future.

COOPERATION AND CIVILIZATION

We all use game theory as unconsciously as sticklebacks or bats. A human society is a group interacting repeatedly. Some interactions pose choices between self and group interest. How often mutual cooperation occurs is a measure of how effectively the society is functioning.

The paramount importance of civilization in human history rests with its role in promoting cooperation. With the discovery of agriculture, people formed permanent settlements. Once people became rooted to cultivated plots of earth, society changed. For the first time, people had neighbors, fellow beings that they would have dealings with again and again. A person who cheated his neighbor could not expect cooperation in the future. A person who cheated all his neighbors would be an outcast. With crops in the ground or an urban business with a stock of good will, it was no longer so easy to pick up and move on. For most people, most of the time, it was easier to cooperate.

Many of the trappings of civilization promote cooperation. Inventions such as names, language, credit cards, license plates, and ID cards help connect a person with his past behavior. That permits the use of conditional strategies. Most laws proscribe defection of various kinds. In their 1957 book, Luce and Raiffa observed that "some hold the view that one essential role of government is to declare that the rules of certain social 'games' must be changed whenever it is inherent in the game situation that the players, in pursuing their own ends, will be forced into a socially undesirable position."

Human history is not one of ever-increasing cooperation, though. Game theory may help to understand that, too. TIT FOR TAT is not the only conditional strategy that is evolutionarily stable or nearly so. Once entrenched, other strategies can be highly stable.

Game theorists Steve Rytina and David L. Morgan investigated the role of *labels*. A label is any category that can be used to distinguish players. In human societies, it may be gender, race, social class, nationality, club membership, union membership, or other attributes.

Imagine a society divided into two groups, the blues and the reds. Nearly everyone in the society follows a strategy that can be called "DISCRIMINATORY TIT FOR TAT" (DTFT). This strategy is just like TIT FOR TAT *except* when dealing with someone of a different color group. Then you always defect.

When two reds interact for the first time, each cooperates. When two blues with no history interact, both cooperate. But when a red and blue interact, each defects ("because you can't trust those guys").

Rytina and Morgan demonstrated that this arrangement is stable. An individual who tries to play regular, color-blind TIT FOR TAT is worse off than one who conforms. Suppose a red and blue interact for the first time, and the blue contemplates cooperating (as in regular TIT FOR TAT). The red player, however, is almost certainly playing DTFT and will defect. The blue player will get the sucker payoff and do less well than a conformist playing DTFT.

This does not mean that DTFT is more successful than TIT FOR TAT would be *if* everyone played TIT FOR TAT. It's not. Every time players of different colors interact they end up with the punishment rather than the reward. But DTFT is stable once entrenched because it punishes individual efforts to establish TIT FOR TAT.

The label that is in the minority is hurt more by DTFT than the majority. If reds are greatly in the majority, then most of a red player's interactions will be with other reds. In that case, DTFT is not much different from TIT FOR TAT. Only in a few cases will an interaction be with a blue. But blues, being in the minority, will interact frequently with reds, and in each case will get the punishment payoff. In the limiting case of an arbitrarily small minority, its members will almost always receive the punishment payoff, while the majority almost always receives the reward.

This provides a game-theoretic rationale of separatist movements. Such diverse phenomena as "Chinatowns" and ghettos, India's partition into Moslem and Hindu states, the Pilgrims' founding of the Massachusetts Bay Colony, the separatism of Marcus Garvey and Black Muslims, and the Mormons' founding of Utah all have or had the effect of limiting a minority's interaction with outsiders distrustful of them.

TIT FOR TAT IN THE REAL WORLD

Many have expressed hope that Axelrod's findings might be applied to human conflicts. One would like to think that statesmen and military leaders would take a course in "practical TIT FOR TAT" and suddenly much of the world's problems would be solved.

Axelrod himself downplays the idea. When I asked him if he thought his findings could be translated into advice for statesmen, he insisted that wasn't the goal. "I think the goal is to help people see things more clearly, which is different. The value of any formal model, including game theory, is that you can see some of the principles that are operating more clearly than you could without the model. But it's only some of the principles. You have to leave off a lot of things, some of which are bound to be important."

Part of the problem with advising anyone to start using TIT FOR TAT in foreign relations is that, in a sense, most reasonable people already do it without knowing it. Responsible leaders don't start trouble, and are provocable. The practical difficulty is not so much in knowing when to cooperate or defect but to decide what is going on. In the real world, it is not always obvious whether someone has acted cooperatively or defected. Actions can fall somewhere between the two extremes, and it is frequently unclear what one's adversary has done. When one cannot tell what the other player has done, it is impossible to use any conditional strategy.

Since the 1950s, the United States and the Soviet Union have used a "tit-for-tat" policy—often called by that name, which predates Axelrod's studies—in granting travel permits to citizens of one nation resident in the other. This appears to be a genuine case of a TIT FOR TAT strategy evolving spontaneously. In 1990, after a American diplomat in Leningrad was denied permission to travel to Lithuania, the U.S. State Department revoked a permit for Gennady Zolotov, Soviet deputy consul general in San Francisco, to travel to give an unrelated and uncontroversial speech at a small college in Nevada. State Department spokesman Chuck Steiner explained, "It wasn't retaliation. It was just an in-kind response. They denied our request, so we denied theirs. It's been a long-held rule between the two countries."

In an uncertain world, TIT FOR TAT-like strategies may be as much a part of the problem as the solution. It is an all too familiar phenomenon of real conflicts that both sides claim the other started it and that they were just giving a tit for a tat. Conflicts escalate mutually. A war of words leads to a war of gunfire and then of air raids. Each side can truthfully cast the other as the side to cross the threshold of war provided it gets to decide where that threshold lies. Bertrand Russell claimed that there was only one war in the history of the world for which people knew the cause: the Trojan War. That was over a beautiful woman; the wars since have lacked rational explanation and have produced nothing, Russell said. In Axelrod's abstract game, there is never any question about who was first to defect, and in this sense it is unrealistic.

Writing in *World Politics* (October 1985) Stephen Van Evera considered whether TIT FOR TAT or a similar strategy might have prevented World War I. He concluded it could not. He said:

> *. . . Tit-for-Tat strategies require that both sides believe essentially the same history; otherwise the players may be locked into an endless echo of retaliations as each side punishes the other's latest "unprovoked" transgression. Because states seldom believe the same history, however, the utility of Tit-for-Tat strategies is severely limited in international affairs. Strategies to promote international cooperation through reciprocity may therefore require parallel action to control the chauvinist mythmaking that often distorts a nation's view of its past. . . .*
>
> *In sum, because conditions required for successful application of a Tit-for-Tat strategy were missing in 1914, Europe was an infertile ground for Tit-for-Tat strategies. These conditions are often absent in international affairs; the syndromes of 1914 were merely pronounced varieties of common national maladies. It follows that we will fail to foster cooperation, and may create greater conflict, if we rely on Tit-for-Tat strategies without first establishing the conditions required for their success.*

How much is game theory presently used in diplomacy? The answer appears to be very little. Axelrod speculated that "I think you can say that [Thomas] Schelling's work is well known and was probably helpful in establishing some of the ideas we have on arms control. But at the very top, you probably cannot find a secretary of state who can tell you what a prisoner's dilemma is."

Axelrod finds game theory's influence more diffuse: "Some of the ideas of game theory are in the public domain very much now, so that somebody can be influenced by them. I think everybody really does know what a non-zero-sum game is. You can use that term in *Newsweek* and not even explain it anymore. Just that is a major intellectual advance because we're so prone to think in zero-sum terms."

13

THE DOLLAR AUCTION

After long days working on the bomb, the scientists of Los Alamos relaxed at dinner parties. The conversation occasionally turned to life on other planets. One night Enrico Fermi asked a question that has never been answered. If there is intelligent life on other planets, where are they? Why don't we detect any sign of them?

Von Neumann's answer is reported in Lewis Strauss's *Men and Decisions* (1962). Strauss says that, shortly after Hiroshima, "Von Neumann had made the semiserious observation that the appearance in the heavens of super-novae—those mysterious stars which suddenly are born in great brilliance and, as quickly, become celestial cinders—could be evidence that sentient beings in other planetary systems had reached the point in their scientific knowledge where we now stand, and, having failed to solve the problem of living together, had at least succeeded in achieving togetherness by cosmic suicide."

It is said that humans are the only earthly creatures aware of their mortality. The generation of the bomb was the first acutely aware of the possible mortality of the human race itself. The power of von Neumann's joking comment rests with the knowledge that it takes only one person to set off catastrophe. Hitler is supposed to have said he wanted a bomb that would destroy the whole world, a story that RAND's Herman Kahn made much of in his analyses of thermonuclear war. In the historic long run, occasional bad leaders and bad decisions are inevitable. The fact that a present-day nuclear war might not quite exterminate the race (it wouldn't release the energy of a supernova, anyway) is beside the point. Von Neumann recognized that the power of weaponry was growing exponentially. In a few generations, a span of time that is but a moment in the history of the earth, weapons rivaling the stars might be possible.

Possibly all civilizations contemplate total war, narrowly avert disaster a number of times, and succeed at disarmament for a while. Then comes one last crisis in which the voices of collective reason are too weak to forestall planetwide holocaust—and that's the reason why we can't detect any radio transmissions from intelligent beings out there.

In a slightly less pessimistic vein, Douglas Hofstadter suggested that there may be two types of intelligent societies in the universe: Type I societies whose members cooperate in a one-shot prisoner's dilemma and Type II societies whose members defect. Eventually, Hofstadter supposed, the Type II societies blow themselves up.

Even this faint optimism is open to question. Natural selection presumably operates much the same way anywhere in the universe. The volatile mixture of cooperation and defection we see on earth may recur on other worlds. If natural selection favors beings who defect in one-shot prisoner's dilemmas, then *all* intelligent species would be genetically "programmed" to do so.

Hofstadter is perhaps talking about a broader kind of natural selection. It's conceivable that natural selection produces only Type II societies, and that, once they reach the state of global technological crisis, they must find a way to transform themselves into a Type I society or die. The question is whether any succeed.

ESCALATION

"I don't think any weapon can be too large," von Neumann once told Oppenheimer, speaking of the hydrogen bomb. The history of warfare is one of escalation to ever more deadly weapons that neither side says it wants. The medieval crossbow, capable of piercing armor, was judged such a dreadful weapon that medieval kingdoms petitioned the Church to outlaw it. Alfred Nobel thought dynamite, so much more powerful than gunpowder, would make war too terrible to contemplate and usher in an age of peace. In their own ways, most of the makers of the bomb, spanning the political spectrum from Oppenheimer to Teller, thought that it would lead to world government and an end to war.

Not only are unprecedentedly terrible weapons used, but they generally spur arms races. The homily that "history repeats itself" is probably nowhere truer than in military history.

In the late 1800s, Britain and France started building battleships to protect themselves from each other. Soon Germany noticed that it was lagging behind. Around the turn of the century, the Kaiser doubled the size of the German fleet. Germany insisted this was for defense, not offense. The British became alarmed at what they now saw as the German threat. Rather than simply build more battleships, they initiated a crash program to produce a more powerful ship. In 1907 the British Admiralty unveiled the Dreadnought. It was faster, better armed, and had more firepower than any previous battleship. Admiral Fisher boasted that one Dreadnought could sink the entire German Navy.

The Germans had no choice but to build their own Dreadnought. Once both nations had Dreadnoughts, the balance of power again hinged on numerical superiority. Britain and Germany embarked on a race to produce the most Dreadnoughts. Germany won the race (which continued until the outbreak of World War I), so Britain actually ended up being less secure. "By adopting the Dreadnought policy we had simply forfeited our enormous preponderance over the prospective enemy without gaining anything in return," complained H. C. Bywater in *Navies and Nations* (1927). ". . . By 1914 the strength of the British Navy as against the German had declined by 40 to 50 per cent. Such were the immediate fruits of the Dreadnought policy."

In many respects the atomic bomb is just the latest chapter in this story. The bomb has cost a lot of money, and has not, in the long run, made anyone more secure. The comforting thought that history repeats itself is spoiled by the fact that it doesn't repeat exactly. Hydrogen bombs are far more terrible than Dreadnoughts or crossbows.

On November 21, 1951, von Neumann wrote a letter to Lewis Strauss critiquing an article by L. F. Richardson on the causes of war that had appeared in the prestigious journal *Nature.* Von Neumann wrote:

> *I feel in sympathy with one of the qualitative ideas that he expresses: Namely, that the preliminaries of war are to some extent a mutually self-excitatory process, where the actions of either side stimulate the actions of the other side. These then react back on the first side and cause him to go further than he did "one round earlier," etc. In other words: This is a relationship of two organizations, where each one must systematically interpret the other's reactions to his aggression as further aggression, and this, after several rounds of amplification, finally leads to "total" conflict.*
>
> *For this reason, also, as the conflict's 'foreplay' progresses, the original aggression, and its motivation, become increasingly ob-*

scured. This would be more comparable to certain emotional and neurotic relationships in ordinary human life than to the more rational forms of antagonism that occur there. As you know, I also believe, that the mere observation that a conflict has some similarity with an emotion or with a neurosis, does not imply per se that the conflict can be resolved—and I think, in particular, that the US-USSR conflict will probably lead to an armed "total" collision, and that a maximum rate of armament is therefore imperative.

Von Neumann may have underestimated game theory's ability to represent "neurotic," irrational acts. The "dollar auction" is a game of escalation and outrageous behavior. It may appear to be the strangest, least realistic game we have encountered. In a deeper sense, the dollar auction is the most faithful allegory of our nuclear age yet. More than any of the dilemmas so far, the dollar auction demonstrates the impotence of game theory in solving certain types of human problems.

SHUBIK'S DOLLAR AUCTION

In their free time, Martin Shubik and colleagues at RAND and Princeton tried to devise new and unusual games. According to Shubik, the central question was, "Can we get certain pathological phenomena as well-defined games?" They wanted games you could actually play. "I don't believe any game that can't be played as a parlor game," Shubik told me.

In 1950, Shubik, John Nash, Lloyd Shapley, and Melvin Hausner invented a game called "so long sucker." This is a vicious game, played with poker chips, where players have to forge alliances with other players but usually have to betray them to win. When tried out at parties, people took the game seriously. ("We had married couples going home in separate cabs," Shubik recalls.)

Shubik posed the question of whether it was possible to incorporate *addiction* in a game. This question lead to the dollar auction. Shubik is uncertain who thought of the game first or whether it was a collaboration. In any case, Shubik published it in 1971 and is generally credited as the game's inventor.

In his 1971 paper, Shubik describes the dollar auction as an "extremely simple, highly amusing and instructive parlor game." A dollar bill is auctioned with these two rules:

1. (As in any auction) the dollar bill goes to the highest bidder, who pays whatever the high bid was. Each new bid has to be higher than the current high bid, and the game ends when there is no new bid within a specified time limit.

2. (Unlike at Sotheby's!) the *second-highest* bidder also has to pay the amount of his last bid—and gets *nothing* in return. You really don't want to be the second-highest bidder.

Shubik wrote, "A large crowd is desirable. Furthermore, experience has indicated that the best time is during a party when spirits are high and the propensity to calculate does not settle in until at least two bids have been made."

Shubik's two rules swiftly lead to madness. "Do I hear 10 cents?" asks the auctioneer—"5 cents?"

Well, it's a dollar bill, and anyone can have it for a penny. So someone says 1 cent. The auctioneer accepts the bid. Now anyone can have the dollar bill for 2 cents. That's still better than the rate Chase Manhattan gives you, so someone says 2 cents. It would be crazy not to.

The second bid puts the first bidder in the uncomfortable position of being the second-highest bidder. Should the bidding stop now, he would be charged 1 cent for nothing. So this person has particular reason to make a new bid—"3 cents." And so on . . .

Maybe you're way ahead of me. You might think that the bill will finally go for the full price of $1.00—a sad comment on greed, that no one got a bargain. If so, you'd be way too optimistic.

Eventually someone does bid $1.00. That leaves someone else with a second-highest bid of 99 cents or less. If the bidding stops at $1.00, the underbidder is in the hole for as much as 99 cents. So this person has incentive to bid $1.01 for the dollar bill. Provided he wins, he would be out only a penny (for paying $1.01 for a dollar bill). That's better than losing 99 cents.

That leads the $1.00 bidder to top *that* bid. Shubik wrote, "There is a pause and hesitation in the group as the bid goes through the one dollar barrier. From then on, there is a duel with bursts of speed until tension builds, bidding then slows and finally peters out."

No matter what the stage of the bidding, the second-highest bidder can improve his position by almost a dollar by barely topping the current high bid. Yet the predicament of the second-highest bidder gets worse and worse! This peculiar game leads to a bad case of buyer's remorse. The highest bidder pays far more than a dollar for a dollar, and the second-highest bidder pays far more than a dollar for *nothing*.

Computer scientist Marvin Minsky learned of the game and popularized it at MIT. Shubik reported: "Experience with the game has shown that it is possible to 'sell' a dollar bill for considerably more than a dollar. A total of payments between three and five dollars is not uncommon." Possibly W. C. Fields said it best: "If at first you don't succeed, try, try again. Then quit. No use being a damn fool about it."

Shubik's dollar auction demonstrates the difficulty of using von Neumann and Morgenstern's game theory in certain situations. The dollar auction game is conceptually simple and contains no surprise features or hidden information. It ought to be a "textbook case" of game theory.

It ought to be a profitable game, too. The game dangles a potential profit of up to a dollar in front of the bidders, and that profit is no illusion. Besides, no one is forced to make a bid. Surely a rational player can't lose. The players who bid up a dollar to many times its value must be acting "irrationally."

It is more difficult to decide where they go wrong. Maybe the problem is that there is no obvious place to draw the line between a rational bid and an irrational one. Shubik wrote of the dollar auction that "a game theory analysis alone will probably never be adequate to explain such a process."

DOLLAR AUCTIONS IN REAL LIFE

Possibly the dollar auction strikes you as nonsense. It's a lot different from a real auction. Don't think about auctions, then. One way to recognize a dollar auction in real life is that it inspires certain figures of speech: "throwing good money after bad"; persevering "so that it all won't have been in vain" or because there is no way to quit and "save face"; having "too much invested to quit."

Have you ever called a busy company long distance and been put on hold for a long time? You can hang up, in which case you've made an expensive call for nothing. Or you can stay on the line, paying more with each passing minute, and with no guarantee that anyone will take the call. It's a true dilemma because no simplistic solution makes sense. Provided you really have to speak to someone at the company and there is no less-busy time to call, you can't categorically resolve to hang up the instant you're put on hold. It's equally ridiculous to say you'll stay on the line no matter how long it takes. There could be some problem with the switchboard and you'd never get connected. It's difficult to decide just how long you should wait, though.

At crowded amusement parks, people end up waiting in line an hour or more for a ride that lasts a few seconds. Sometimes you can't believe you waited in line so long for so little. The reason is the "human-engineered" serpentine queues that prevent patrons from seeing how long the line is. You patiently work your way up to a certain point, then turn a corner and see a whole new part of the line. By the time you appreciate just how long the line is, you've already invested too much time to give up.

Allan I. Teger found that dollar-auction-like situations are frequently created or exploited for profit. In his 1980 book, *Too Much Invested to Quit,* Teger notes, "When we are watching a movie on television only to discover that the movie is poor, we are reluctant to turn it off, saying that we have watched it so long that we might just as well continue to see how it ends. . . . The television stations know that we are reluctant to turn the movie off once we have begun to watch it, so they will often increase the length and frequency of commercials as the movie progresses. Seldom do we turn off the movie with only 20 minutes remaining, even if the commercials are coming at the rate of one every five minutes."

Strikes that threaten to ruin both labor and management have much in common with the dollar auction. Each side wants to stick it out a little longer; if they give in now all the lost wages or lost profits would just be money down the drain. The dollar auction resembles architectural design competitions (architects invest their own time and effort designing a prestigious new building, but only the winner gets the commission) and patent races (competing firms invest research and development funds on a new product, but only the first to patent it makes any money). Repairing an old car—playing a few more hands of cards to recoup losses—waiting for the bus a few minutes more before giving up and hailing a taxi—staying in a bad job or bad marriage: all are dollar auctions.

As we've seen, these game-theoretic dilemmas have a way of being discovered at an appropriate moment in history. The conventional perception of the Vietnam conflict—particularly the psychology popularly imputed to Presidents Johnson and Nixon—is pure dollar auction. "Winning," in the sense of improving American interests to a degree that might justify the lives lost and money spent, was scarcely possible. The main agenda was to push a little harder and get a nominal victory—"peace with honor," so that our dead will not have died in vain. Shubik recognizes the Vietnam war as an "exquisite example" of a dollar auction but doesn't recall it being an inspiration for the game. He believes the game predated his 1971 publication by some time, which might have put its genesis before the late stages of the war.

More recently, the Persian Gulf war was a rich source of dollar-auction rhetoric. In a January 1991 speech to his troops on the southern front, Iraqi President Saddam Hussein "declared . . . that Iraq's material losses are already so great that he must now fight to the end," according to a *Los Angeles Times* story (January 28, 1991).

Shubik cites Hussein's position as a particularly troubling case of the myopia of real leaders in such situations. Both sides in the Vietnam conflict could with some plausibility entertain hopes of holding out and winning. The Iraq conflict was far more lopsided. Iraq fought with a technologically backward army a fraction of the size of the UN coalition's forces. Iraq's crushing defeat was predictable, seemingly, to everyone but Hussein. It's easy to dismiss Hussein as a lunatic. Unfortunately, this brand of lunacy is a prevalent one. People aren't always good at predicting how others will react to their actions. It's easy to blind oneself to the consequences.

Dollar auction-like conflicts occur in the animal world. Territorial struggles between animals of the same species rarely lead to fights to the death. Sometimes they are mere "wars of attrition," where combatants face off and make threatening gestures. Finally one animal gets tired and leaves, conceding defeat. The animal willing to hold its ground the longest wins. The only "price" the animals pay is time (time that might be used to hunt food, mate, or care for offspring). Both animals pay the same price; only the one willing to hold out longer wins the dispute.

The dollar auction bears some resemblance to an iterated prisoner's dilemma. Topping a bid is defecting, since it betters the bidder's short-term individual position while hurting the common good. Every new bid chips away at the potential profit. The usual debacle is the result of repeated defection on both sides.

Sometimes the dollar auction is a better model than the iterated prisoner's dilemma for some of the conflicts conventionally treated as such. Escalation, and the possibility of ruin for both sides, is characteristic of arms races. The "winner," the nation that builds the biggest and most bombs, wins a measure of security. The "loser," however, is not only less secure for it; it does not get its "wasted" defense budget refunded, either. Consequently, the second-strongest superpower is tempted to spend a little more money to "close the missile gap."

The dollar auction hints at the difficulty of applying strategies like TIT FOR TAT. Each bidder is echoing the "defection" of the other! To stop bidding is to allow oneself to be exploited.

You might suppose that the problem is that the bidders aren't "nice" in Axelrod's sense. Lay the blame at the door of the first person to defect—that is, the first bidder. But how can you criticize the person who makes the first bid? If no one bids, the ninety-nine-cent profit is wasted.

Many conflicts start this way, with a justifiable action that, in retrospect, becomes the first "defection" in an escalating dilemma. The nuclear arms race between the United States and the Soviet Union started when the United States made a bomb to defeat Adolf Hitler, an unquestioned belligerent who was working on his own atomic bomb. It's hard to find fault with that. Jacob Bronowski, one of the scientists who worked on the bomb, put it this way *(The Listener,* July 1, 1954):

> *The scale of the damage of Nagasaki drained the blood from my heart then (i.e., in autumn, 1945), and does so now when I speak of it. For three miles my road lay through a desert which man had made in a second. Now, nine years later, the hydrogen bomb is ready to dwarf this scale, and to turn each mile of destruction into ten miles, and citizens and scientists stare at one another and ask: "How did we blunder into this nightmare?"*
>
> *I put this first as a question of history, because the history of this is known to few people. The fission of uranium was discovered by two German scientists a year before the war. Within a few months, it was reported that Germany had forbidden the export of uranium from the mines of Czechoslovakia, which she had just annexed. Scientists on the continent, in England and America, asked themselves whether the secret weapon on which the Germans were said to be working was an atomic bomb. . . . The monopoly of such a bomb would give Hitler instant victory, and make him master of Europe and the world. The scientists knew the scale of what they feared very well: they feared first desolation and then slavery. . . . Einstein had been a pacifist all his life,*

and he did not easily put his conscience on one side. But it seemed clear to him that no one was free to keep this knowledge to himself. . . . On August 2, 1939, a month before Hitler invaded Poland, Einstein wrote to President Roosevelt to tell him that he thought an atomic bomb might be made, and he feared that the Germans were trying to make one. This is how it came about that, later in the war, scientists worked together in England, in Canada and America, to make the atomic bomb. They hated war no less than the layman does—no less than the soldier does . . . The atomic scientists believed that they were in a race against Germany whose outcome might decide the war even in its last weeks.

Once the bomb was made, there was no way of unmaking it. The U.S. bomb led the Soviets and other nations to make their own, which led the United States to make a hydrogen bomb, and then the Soviets to make one, and then for both sides to make a lot of them. Where do you draw the line?

STRATEGIES

In Shubik's trials, bidders acted spontaneously without examining their actions nearly as exhaustively as has been done in dozens of scholarly papers since then. What should you do in a dollar auction? Should you bid at all?

As in many experiments, not everyone takes a real dollar auction seriously. Some people inflict losses on the other bidders just for the hell of it. Others incline toward a "gentlemen's understanding" that one person should bid a cent and the others should pass, letting him have it. To do justice to Shubik's intention, we have to assume that each bidder is interested only in maximizing his personal gain (or minimizing loss, if need be) and that the monetary amounts are meaningful to the bidders.

Take the case where there are exactly two bidders. The minimum bid, and the minimum increment, is a cent. Each of the two players successively take turns bidding. Each new bid must top the previous one. Otherwise, the player "passes," in which case the other player wins the dollar. Suppose you're the first to bid. Let's look at the possibilities in turn:

Bid 1 cent. This is the minimum bid. Should the other player be nice enough to pass, it would give you the maximum profit (99 cents). It's also the "safest" bid in that you can lose only a penny if the player tops you and you make no follow-up bids. Bidding 1 cent means maximum profit and minimum risk—what more do you want?

Unfortunately, the other player has every incentive to top your bid of 1 cent. It is far from clear what the ultimate outcome of this opening bid would be.

Bid 2 cents to 98 cents. Any bid in this range gives you some profit but also lets the other player top it and make a profit. It's hard to say what would happen.

Bid 99 cents. This is the maximum bid which could possibly lead to a profit. Provided no fractional cents are allowed, the other player's options are to bid at least $1.00, or to pass. The other player would be a fool to bid more than $1.00, for that guarantees a loss (an unnecessary loss, for he has the option of passing and breaking even). And he has no incentive to bid $1.00. He would also break even if the bidding stopped there, but he takes the risk of losing the dollar if you subsequently top his bid.

A bid of 99 cents will win you a profit of 1 cent provided the other player acts conservatively and with no animosity toward you.

Bid $1.00. This nihilist move doesn't make much sense. It immediately wipes out any expectation of profit on anybody's part. It preempts even the slight uncertainty about the other bidder's response noted in the 99-cent case. Now the minimum follow-up bid is $1.01, which guarantees the other bidder a loss. If the other player is interested in maximizing his outcome, he has to pass.

Bid more than $1.00. Well, that's just silly.

There is one more strategy. It is to—

Pass (even though nothing has been bid). You can't win anything this way. You don't risk anything, either. If you bid a dollar, you accept the slight risk that the other player will irrationally top it and you will be out the dollar. So passing is definitely better than bidding $1.00, for what it's worth. Why accept any risk when you don't have to?

If you pass, that leaves the other player to name his price for the dollar. He will bid 1 cent and get the dollar for a profit of 99 cents. At least you don't spoil it for the other player.

Of the possible opening bids, only 99 cents promises a profit, and that the rock-bottom minimum. You may not even feel that the 1-cent profit makes up for the slight risk of losing the 99 cents if the other player is irrational.

It strikes most people as wrong that the potential profit of 99 cents should go to waste. This profit accrues (to the other player, unfortunately) only if you pass.

Coming full circle, let's reexamine the first strategy, of the first player bidding 1 cent. Why shouldn't the *second* player be sensible enough to pass and let the first take his profit? Look at it this way: *No one can guarantee himself a profit by topping a previous bid.* Would not a rational player realize it's a sucker's game and refuse to top any existing bid?

The potentially profitable topping bids are in the range of 2 cents to 99 cents (a bid of 1 cent has to be a first bid, not a topping bid). The profit is insecure unless the other player will have no incentive to top the bid himself. But the first player has already committed at least 1 cent himself. Your topping bid puts the other bidder at least 1 cent in the hole. Consequently, he now has an incentive to break even. He can be expected to top your "profitable" bid. This would leave you in the hole, and in the thick of a mad cycle of escalation.

RATIONAL BIDDING

The dollar auction is subject to the analysis of game theory. There is a "rational" solution, provided the bidders have known finite bankrolls. With a fixed bankroll, the bidding can go on only so long. Therefore it is possible to tally all conceivable sequences of bids. Once all this bookkeeping is done, it is possible to start at the final bids and backtrack. In separate papers, Barry O'Neill and Wolfgang Leininger did this.

Leininger's 1989 paper considered the case where there is a continuous range of bids (fractional cents allowed). Only slight adjustment is necessary when bids must be integral multiples of a cent.

Let's take a concrete situation. The "stake" is, as usual, $1.00 and each bidder has a "bankroll" of, say, $1.72. According to Leininger, the correct first bid is the *remainder* when the stake is divided by the bankroll—$1.72 divided by $1.00 leaves a remainder of 72 cents. That's what the first player should bid.

At first sight, this may appear crazy. But look: a bid of 72 cents leaves the second player tempted to make a bid in the range of 72 cents to $1.00, the range of profit. It is also a range of vulnerability for the first player, for any bid topping his costs him the stake and puts him out 72 cents to boot. But if the second player does top the first bid for a profit, the first player is willing and able to bid his whole bankroll of $1.72 (and end the bidding right there—the second player doesn't have enough money to top it). This threat is a credible one because making good on it won't cost him a cent. If the first player does nothing after his bid is topped, he loses his initial bid (a 72-cent loss). If he bids his whole bankroll, he pays $1.72 but is guaranteed to recover a dollar (a 72-cent loss). He's no worse off doing this.

In effect, the first player's threat by bidding 72 cents is this: "Drop out and let me take this 28-cent profit. If you don't, I'll make sure you lose by bidding my whole bankroll—it's no skin off my nose."

Not only can the first player make good on this threat, but there is nothing the second player can do to forestall it. He can't bet less than 72 cents since a new bid has to be higher than the existing one. He can't do what he'd like, bid in the zone of profit, because of the threat. If he's reasonable, he can't even say, "Two can play at that game" and beat the first player to the punch by bidding his whole bankroll. Then he would win all right—a $1.72 dollar bill, which is no bargain. The second player is better off settling for nothing.

The "rational" first bid would be 72 cents whether the bidders' bankrolls were $1.72, $2.72, or $1,000,000.72. The bidders can "chop off" a dollar from the bankroll in each round of bidding, so that it is the remainder that determines the strategy. Suppose that the bankroll *is* $1,000,000.72. When the first player bids 72 cents, the second will want to bid in the range of 72 cents to $1.00 but is deterred by the threat of the first player bidding $1.72. If however the second player dares to make a profitable bid (we assume he will never "cut off his nose to spite his face" and make an unprofitable bid), the first player retaliates by bidding $1.72 (which doesn't hurt him).

Now that the second player is in the hole for the amount of his first bid (72 cents to $1.00) he has incentive to make a new bid as high as $1.72 to $2.00. He should really accept his losses . . . but if he is so foolhardy as to make a new bid, then the first player is ready and willing to top it. Once the second player makes a new bid, the first player is $1.72 in the hole. This gives him an incentive to bid as high as $2.72 to recapture the lead.

As the bidding continues, each bidder is increasingly worse off, but each has incentive to continue in the short term . . . until the first player hits rock bottom in the form of the bidding limit. Then he wins —make that "wins"—because he's just paid $1,000,000.72 for a dollar bill. Nevertheless, this remote contingency is the first bidder's ace in the hole. He knows that, if the second player insists on continuing the bidding, he (the first player) will achieve a victory, however Pyrrhic.

Now we can see why the first player has to bid 72 cents. He might, of course, want to bid less than that in the belief it will increase his profit. If he does, however, the second player can immediately bid 72 cents and then follow the "first-player" strategy outlined above. Then the second player will end up victorious.

The situation is completely different, and simpler, when one player has more money to spend than the other. The player with more money —even a penny more—can always outbid the other if forced to do so. Consequently, the richer player can (and should) make an opening bid of 1 cent, confident that his poorer but rational adversary will realize the folly of a bidding war.

WHERE GAME THEORY FAILS

As the Persian Gulf war shows, it doesn't always work out that simply. "After high school, there is no such thing as algebra," the joke goes. Mathematics, including game theory, is often far-removed from real life.

I think Shubik's moral is that game theory's recommendation is beside the point, insofar as real dollar auctions are concerned. The marvelous crystalline logic wouldn't begin to apply in a real auction or a real geopolitical crisis.

Even were all the relevant facts known, a dollar auction is much like a game of chess. There is a rational and correct way to play chess, but you would be very wrong to assume that your opponent is playing this way. No one can look far ahead enough to see game theory's solution to chess. Few are likely to look that far ahead in a dollar auction.

This is only part of the problem. In real dollar auctions, there is uncertainty about how much players have to bid. In actual tests of the dollar auction, bidders can only guess at how much money their opponents have on them—at least until a late stage where someone empties a wallet to see how much he has. The backward analysis of game theory requires that the bidding limits be known from the outset. If they are indefinite, there is nothing to work from.

To apply game theory to a real dollar auction such as an arms race, you would have to know *exactly* how much nations are willing and able to spend on defense. No one can do more than guess at this. It's overly simplistic even to try to assign a single number to it. The money must be spent over a number of years, and public opinion is flexible. A nation that resolves to beef up its defenses now may change its mind after a few years of large military budgets and high taxes.

To be sure, no estimate or measurement of anything is precise. In many branches of science, exact measurement is not essential to useful application of theory. We don't know the mass of the earth and moon with infinite precision, but that doesn't prevent us from navigating rockets to the moon. Small uncertainties about the masses of celestial bodies lead to proportionately small uncertainties about the trajectory of a rocket.

Game theory's solution to the dollar auction takes an unusual form. The rational course of actions depends on knowing with certainty which of two inexact quantities is bigger, or (if equal) the *remainder* when one inexact quantity is divided by yet another. Given the typical degree of uncertainty about bankroll, the uncertainty about the rational bid will not merely be in proportion to the former uncertainties. The uncertainty will be *total.* It's a case of "garbage in, garbage out."

The previous chapters made a distinction between von Neumann's all-inclusive strategies and limited or provisionary strategies. The problem with the dollar auction is not simply that we don't know the limits to the game. The limits of an iterated prisoner's dilemma may be unknown, but there are good limited strategies. Ignorance is bliss: TIT FOR TAT is better than Nash's all-inclusive strategy of constant defection. In the dollar auction, however, there is no good limited strategy.

You find yourself in a dollar auction with an unknown bidder(s). You don't know how much money you have or the other(s) have. The auction can go on for several bids (or a pass) at least; possibly longer. What you want is a strategy that works pretty well in this state of ignorance. It would allow you to make a profit and convince the other bidder(s) to drop out, or, failing that, minimize any losses.

As we've seen, that appears to be impossible. Prospects for a dollar auction are much more pessimistic than an iterated prisoner's dilemma. In a realistic situation, against players who cannot be expected to see more than a few bids ahead, there is really nothing to do but refuse to play. But you can hardly call that a good way to play. There *is* a prize for the taking.

The greater tragedy is that we often do not realize we are in a dollar auction until we have already made several bids. The crux of the matter is whether to make another bid when one is the second-highest bidder. By not bidding, and accepting the loss, you end the cycle of escalation. But why should *you* be the one to take the loss? Why not the other guy, who just raised your previous bid? If the player who ends the escalation is the more "rational" one, the dollar auction penalizes rationality.

In practice, Teger maintained, a cycle of escalation is broken by a pretext that allows one or both parties to save face. In a protracted strike, both sides may realize they are in a no-win situation but are afraid of being thought foolish if they give up. They are likely then to seize on any reason for ending the deadlock. One side may suddenly announce that the real issue is X, something that they know the other side will readily concede and which was never at issue. The other side will agree to X, and the strike will end. It's a matter of personalities, group psychology, and luck, not game theory; rationality has nothing to do with it.

THE LARGEST-NUMBER GAME

Another caricature of escalation is a game devised by Douglas Hofstadter. It is known as the "luring lottery" or "largest-number game."

Many contests allow unlimited entries, and most of us have at some time daydreamed about sending in millions of entries to better the chance of winning. The largest-number game is such a contest, one that costs nothing to enter and allows a person to submit an unlimited number of entries.[1] Each contestant must act alone. The rules strictly

1. Stories of misguided people who try to win state lottery jackpots by buying a huge number of tickets occur in the news now and then. The most publicized case of massive multiple entries in a *free* lottery occurred in 1974. Caltech students Steve Klein, Dave Novikoff, and Barry Megdal submitted about 1.1 million entries on 3 by 5 pieces of paper to a McDonald's restaurant sweepstakes. The Caltech entries amounted to about one fifth of the total entries, winning the students a station wagon, $3,000 cash, and about $1,500 in food gift certificates. Other contestants were outraged. Many quoted in the press were sure they had been cheated out of something, even though the rules explicitly allowed unlimited entries. "We were disappointed to hear of the sabotage of your efforts by a few 'cuties'!" a man from Orange, California, wrote McDonald's. "Un-

forbid team play, pools, deals, or any kind of communication among contestants.

All entries have an equal chance of winning. On the night of the drawing, the contest's sponsors pick one entry at random and announce the lucky winner. He or she wins up to a million dollars, according to rules of the contest.

When the game's generous sponsors say "unlimited" entries, they mean it. A million entries—a billion entries—a trillion entries—all are perfectly okay. The number must of course be finite, and it must also be a whole number. Just so you understand how it works: suppose that 1 million other people have entered the contest, and all of them—meek souls—have submitted a single entry. Then you submit a million entries yourself. That makes 2 million entries in all. Your chance of winning would be 1 million/2 million, or 50 percent. But if someone else came along and submitted 8 million more entries just before the deadline, there would be 10 million entries total, and your chance of winning would drop to 1 million/10 million, or 10 percent. The person with the 8 million entries would then stand an 80 percent chance of winning.

Obviously, it's to your advantage to submit as many entries as possible. You want to submit more than anyone else; ideally, a *lot* more than all the other contestants put together. Of course, everyone wants to do this. There is a practical limit to how many entries anyone could fill out and send in. Besides, the postage would be prohibitive . . .

No problem! The rules say that all you have to do is send in a single 3 by 5-inch card with your name and address and a number indicating how many entries you wish to submit. If you want to submit a quintillion entries, just send in one card with "1,000,000,000,000,000,000" written on it. It's that simple. You don't even have to write out all those zeros. You can use scientific notation, or the names of big numbers (like "googol" or "megiston"), or even high-powered exotica like the Knuth arrow notation. Qualified mathematicians will read each card submitted and make sure everyone gets proper weight in the random drawing. You can even invent your own system for naming big numbers, as long as there is room to describe it and use it on the 3 by 5 card.

There is one catch.

---

doubtedly all giveaway contests in the future will have to be safeguarded in some way from similar tactics by 'punks' such as these." Those sharing this man's feelings are advised to avoid the largest-number game!

A minuscule asterisk on the lottery posters directs your attention to a note in fine print saying that the prize money will be $1 million *divided by the total number of entries received.* It also says that if no entries are received, nothing will be awarded.

Think about that. If a thousand people enter and each submits a single entry, the total entries are 1,000 and the prize money is $1 million/1000, or $1,000. That's nothing to cry in your beer over, but it's no million.

If just one person is so bold as to submit a million entries, the maximum possible prize money plummets to *1 lousy dollar.* That's because there will consequently be at least a million entries (probably many more.) If anyone submits 100 million entries, the prize can be no more than a cent, and if more than 200 million entries are received, the "prize," rounded down to nearest cent, will be nothing at all!

The insidious thing about the largest-number game is the slippery way it pits individual and group interests against one another. Despite the foregoing, the game is *not* a sham. Its sponsors have put a cool $1 million into an escrow account. They are fully prepared to award the full million—provided, of course, that just one entry is received. They are prepared to award any applicable fraction of the million. It would be a shame if *someone* doesn't win *something.*

How would you play this game? Would you enter, and if so, how many entries would you submit?

You'll probably agree that it makes no sense even to think about submitting more than 200 million entries. That unilateral act, on the part of anybody, effectively wipes out the prize. Fine. Extend this reasoning. Maybe you feel that a dollar is the smallest amount worth bothering over. Or you reason that people's time is worth something. Estimate how long it takes to fill out a 3 by 5 card (including the time spent deliberating) and multiply this by a minimum wage or a billing rate. Add in the price of a postage stamp. The result is what it costs to enter the contest. It makes no sense to submit a number of entries that would itself suffice to lower the prize money below the threshold of profit.

Taking this to an extreme, maybe the best plan is to submit just one entry. Maybe everyone else will do the same. Then at least everyone will have a "fair chance"—whatever *that* means in a largest-number game.

Maybe you shouldn't enter the contest at all. Then no one could say you weren't doing your share to keep the prize money high. The thing is, the largest-number game is like all lotteries: if you don't play, you can't win. Worse yet, if *everyone* decided not to play, no one would win.

In the June 1983 issue of *Scientific American,* Douglas Hofstadter announced a largest-number game open to anyone who sent in postcards before midnight, June 30, 1983. The prize was $1 million divided by the total number of entries. The management of *Scientific American* agreed to supply any needed prize money.

The result was predictable. Many of the correspondents sought to *win* rather than to maximize winnings. The game became, in effect, a contest to see who could specify the biggest integer and thereby get his name in the magazine (as "winner" of an all but infinitesimal fraction of a cent). Nine people submitted a googol entries. They had their chances wiped out by the fourteen contestants who submitted a googolplex. Some people filled their postcard with nines in microscopic script (designating a number bigger than a googol but much, much smaller than a googolplex). Others used exponents, factorials, and self-defined operations to specify much bigger numbers, and some crammed the postcard with complex formulas and definitions. Hofstadter couldn't decide which of the latter entries was the biggest, and no one got his name in the magazine. Of course, it didn't matter financially who won. The prize amount would round to zero; and had *Scientific American* issued a check for the *exact* amount, it would have had to employ the same mathematical hieroglyphics the winning contestant had used, and no bank would accept it!

Greed is part and parcel of the largest-number game, but not the incentive to be thought clever by naming the biggest number. In a true largest-number game, the players are motivated only by maximizing personal gain. The real question of how to play the game remains.

No matter what, only one person can win. The best possible outcome of the game, then, is for the full million dollars to be disbursed to the winner. This occurs only when there is a single entry. *If* the contestants were permitted to work out some share-the-wealth scheme and coordinate their actions—which they are not—they would surely arrange to have just one person submit an entry.

The largest-number game is almost a mirror image of the volunteer's dilemma. You want everyone *except* one person to volunteer *not* to enter. The ideal would be to draw straws so that exactly one person gets the go-ahead to enter. This unfortunately requires explicit communication among the players, which is not permitted.

The game appears to be one of pure defection, with no cooperative way to play. That's not so. There *is* a collectively rational way to play the largest-number game (or a volunteer's dilemma). It is a mixed strategy: for each and every person to decide whether to volunteer based on a random event such as a roll of dice.

Each person can roll his own dice independently of the others, without any need of communication. If there are thirty-six players, then each might roll two dice and submit a single entry only if snake eyes came up (a 1 in 36 chance).[2]

This would work just fine. But the mind reels at such "rationality." This isn't what happened in the *Scientific American* lottery, and it is hard to imagine any mass of living, breathing human beings acting as prescribed. The temptation to cheat is superhuman. You roll your pair of dice and it comes up snake eyes—oops, no, the one die hit the edge and now it shows a three! You don't get to enter. Who would know if you helped Dame Fortune along and flipped the die back over? No one! Everyone rolls the dice in the privacy of his home and acts independently. Who would know if you rolled a fair and square four and entered anyway? No one!

This rationalization doesn't change things one bit. If everyone reasons like this, everyone will enter and wipe out the prize. It's like the volunteer's dilemma, only worse. In a volunteer's dilemma, everyone is penalized according to the proportion of defectors among the players. In the largest-number game, even one defector can ruin it for everyone.

So what do you do? Do you go through the almost pointless exercise of calculating odds and rolling dice (knowing full well that most others are doing no such thing, and that some are already filling cards with as many nines as the points of their pencils will permit), or do you fill a card with nines yourself? The only reasonable conclusion is that the largest-number game is a hopeless situation.

2. Actually, it's best to adjust the odds slightly. The trouble with giving each of n persons a $1/n$ chance is that there's still a fairly big chance that no one will get the go-ahead to enter. By the laws of probability, this chance is about 37 percent when n is a few dozen or more. This is balanced by the chance that two or three or more people will get the go-ahead.

A better plan is to allow each person a $1.5/n$ chance of getting the go-ahead. This decreases the chance of no one entering at the expense of increasing somewhat the chance of extra entries.

FEATHER IN A VACUUM

The dollar auction and largest-number game share important features. Shortsighted rationality forces players to subvert the common good. And when thoughtful players try to do the collectively rational thing, they are all too likely to be exploited. These are the same issues faced in a prisoner's dilemma, and once again we find them a recurring motif in other, more complex problems.

Martin Shubik wrote (1970), "The paradox of the Prisoner's Dilemma will never be solved—or has already been solved—because it does not exist." He meant that rational players will defect in a one-shot dilemma, and that the game demonstrates just what it was intended to; namely, that individual interests can overturn the common good. Shubik compared puzzlement over social dilemmas to the average person's surprise at seeing a feather and a lead weight fall at the same speed in a vacuum. Our intuition does not prepare us for either phenomenon, but the "explanation" is simply that common sense is wrong.

A game is just a rectangular matrix filled with numbers—any numbers at all. Because a game can have any rules whatsoever, it's possible to design games that penalize any fixed notion of rationality. This is no more remarkable than the claim that you can design a machine to penalize rationality. Build a steel trap door, have the "victim" fall through it into a maze designed to test rationality, and if he succeeds within a time limit, the machine extracts his wallet.

The paradox lies in the fact that our concept of rationality is not fixed. When one type of "rational" behavior fails, we expect the *truly* rational person to step back, think things over, and come up with a new type of behavior. Given this open-ended definition of rationality, it's hard to accept that rationality ever runs aground. In a one-shot prisoner's dilemma, the type of rationality game theory recognizes leads to mutual defection, and all attempts to come up with alternative types of rationality have failed.

Real-world dilemmas are built of subjective valuations of the welfare of one's self and others. If there's a note of hope, it's that these feelings are flexible. Cold war propaganda painting the "enemy" as a nation of heartless automatons is the sort of thing that predisposes people to prisoner's dilemmas. The ability to see "opponents" as fellow beings frequently transforms a nominal prisoner's dilemma into a much less troublesome game. The only satisfying solution to the prisoner's dilemma is to avoid prisoner's dilemmas.

This is what we've been trying to do all along with laws, ethics, and all the other cooperation-promoting social machinery. Von Neumann was probably right, in that the long-term survival of the human race depends on our devising better ways to promote cooperation than any now in existence. The clock is ticking.

BIBLIOGRAPHY

Abend, Hallett. *Half Slave, Half Free: This Divided World.* Indianapolis: Bobbs-Merrill, 1950.

Axelrod, Robert. *The Evolution of Cooperation.* New York: Basic Books, 1984.

Bascom, William. "African Dilemma Tales: An Introduction." In *African Folklore,* edited by Richard M. Dorson. Bloomington, Ind., and London: Indiana University Press, 1972.

Blair, Clay, Jr. "Passing of a Great Mind." In *Life,* February 25, 1957, 89–90+.

Blumberg, Stanley A., and Gwinn Owens. *Energy and Conflict: The Life and Times of Edward Teller.* New York: G. P. Putnam's Sons, 1976.

Bott, Raoul. "On Topology and Other Things." In *Notices of the American Mathematical Society,* 32 (1985) no. 2, 152–58.

Bronowski, Jacob. *The Ascent of Man.* London: British Broadcasting Corporation, 1973.

Clark, Ronald W. *The Life of Bertrand Russell.* Harmondsworth, England: Penguin Books, 1978.

Courlander, Harold, and George Herzog. *The Cow-Tail Switch, and Other West African Stories.* New York: Henry Holt & Co., 1947.

Cousins, Norman. *Modern Man Is Obsolete.* New York: Viking, 1945.

Davis, Morton D. *Game Theory: A Nontechnical Introduction.* New York: Basic Books, 1970.

Davis, Nuel Pharr. *Lawrence & Oppenheimer.* New York: Simon & Schuster, 1968.

Dawkins, Richard. *The Selfish Gene.* 2d ed. Oxford: Oxford University Press, 1989.

Dickson, Paul. *Think Tanks.* New York: Atheneum, 1971.

Fermi, Laura. *Illustrious Immigrants.* Chicago: University of Chicago Press, 1968.

Flood, Merrill M. "Some Experimental Games." Research Memorandum RM-789. Santa Monica, Calif.: RAND Corporation, 1952.

Goldstine, Herman H. *The Computer from Pascal to von Neumann.* Princeton, N.J.: Princeton University Press, 1972.

Goodchild, Peter. *J. Robert Oppenheimer: Shatterer of Worlds.* Boston: Houghton Mifflin, 1981.

Grafton, Samuel. "Married to a Man Who Believes the Mind Can Move the World." In *Good Housekeeping,* September 1956, 80–81+.

Guyer, Melvin J., and Anatol Rapoport. "A Taxonomy of 2 × 2 Games." In *General Systems* (1966) 11:203–14.

Haldeman, H. R., with Joseph DiMona. *The Ends of Power.* New York: Times Books, 1978.

Halmos, Paul. "The Legend of John von Neumann." In *American Mathematical Monthly* 80, no. 4 (April 1973): 382–94.

Heims, Steve J. *John von Neumann and Norbert Wiener: From Mathematics to the Technologies of Life and Death.* Cambridge, Mass.: MIT Press, 1980.

Hobbes, Thomas. *Leviathan.* New York: Macmillan, 1958.

Hofstadter, Douglas. *Metamagical Themas: Questing for the Essence of Mind and Pattern.* New York: Basic Books, 1985.

Kahn, Herman. *On Escalation: Metaphors and Scenarios.* New York: Praeger, 1965.

———. *On Thermonuclear War.* Princeton, N.J.: Princeton University Press, 1960.

Keohane, Robert O. *After Hegemony: Cooperation and Discord in the World Political Economy.* Princeton, N.J.: Princeton University Press, 1984.

Kraft, Joseph. "RAND: Arsenal for Ideas." In *Harper's,* July 1960.

Luce, R. Duncan, and Howard Raiffa. *Games and Decisions.* New York: John Wiley & Sons, 1957.

Lutzker, Daniel R. "Internationalism as a Predictor of Cooperative Behavior." In *Journal of Conflict Resolution* 4 (1960): 426–30.

———. "Sex Role, Cooperation and Competition in a Two-Person, Non-Zero-Sum Game." In *Journal of Conflict Resolution* 5 (1961): 366–68.

Marshall, George C. Quoted in *Foreign Relations of the United States,* 1948, III:281.

Maynard-Smith, John. *Evolution and the Theory of Games.* Cambridge: Cambridge University Press, 1982.

Milinski, Manfred. "TIT FOR TAT in Sticklebacks and the Evolution of Cooperation." In *Nature* 325 (January 29, 1987): 433–35.

Minas, J. Sayer, Alvin Scodel, David Marlowe, and Harve Rawson. "Some Descriptive Aspects of Two-Person Non-Zero-Sum Games. II." In *Journal of Conflict Resolution,* 4 (1960): 193–97.

Moran, Charles. *Churchill, Taken from the Diaries of Lord Moran.* Boston: Houghton Mifflin, 1966.

Morgenstern, Oskar. *The Question of National Defense.* New York: Random House, 1959.

Morton, Jim. "Juvenile Delinquency Films." In *Re/Search* 10 (1986): 143–45.

Neumann, John von. "Can We Survive Technology?" In *Fortune* (June 1955): 106–8+.

———. "Communication on the Borel Notes." In *Econometrica* 21 (1953): 124–25.

———. *The Computer and the Brain.* New Haven: Yale University Press, 1958.

———, and Oskar Morgenstern. *Theory of Games and Economic Behavior.* Princeton, N.J.: Princeton University Press, 1944.

Newman, James R. *The World of Mathematics.* New York: Simon & Schuster, 1956.

Oye, Kenneth, ed. *Cooperation Under Anarchy.* Princeton, N.J.: Princeton University Press, 1986.

Payne, Robert. *The Great Man: A Portrait of Winston Churchill.* New York: Coward, McCann & Geoghegan, 1974.

Pfau, Richard. *No Sacrifice Too Great: The Life of Lewis C. Strauss.* Charlottesville: University of Virginia Press, 1984.

Rapoport, Anatol. "Experiments with N-Person Social Traps I." In *Journal of Conflict Resolution* 32 (1988): 457–72.

———. *Fights, Games, and Debates.* Ann Arbor: University of Michigan Press, 1960.

———. "The Use and Misuse of Game Theory." In *Scientific American* (December 1962): 108–14+.

Rousseau, Jean Jacques. *A Discourse on Inequality.* Translated by Maurice Cranston. London: Penguin, 1984.

Russell, Bertrand. *The Autobiography of Bertrand Russell.* Boston: Little, Brown, 1967–69.

———. *Common Sense and Nuclear Warfare.* New York, Simon & Schuster, 1959.

———. *Unarmed Victory.* New York: Simon & Schuster, 1963.

Schelling, Thomas C. *The Strategy of Conflict.* Cambridge, Mass.: Harvard University Press, 1960.

Scodel, Alvin, J. Sayer Minas, Philburn Ratoosh, and Milton Lipetz. "Some Descriptive Aspects of Two-Person Non-Zero-Sum Games." In *Journal of Conflict Resolution* 3 (1959): 114–19.

Seckel, Al. "Russell and the Cuban Missile Crisis." In *Russell* (Winter 1984–85): 253–61.

Shepley, James R., and Clay Blair, Jr. *The Hydrogen Bomb.* New York: David McKay, 1954.

Shubik, Martin. "The Dollar Auction Game: A Paradox in Noncooperative Behavior and Escalation." In *Journal of Conflict Resolution* 15 (1971): pp. 545–47.

———. "Game Theory, Behavior, and the Paradox of the Prisoner's Dilemma: Three Solutions." In *Journal of Conflict Resolution* 14 (1970): 181–93.

———, ed. *Game Theory and Related Approaches to Social Behavior: Selections.* New York: John Wiley & Sons, 1964.

Smith, Bruce. *The RAND Corporation.* Cambridge, Mass.: Harvard University Press, 1966.

Stern, Philip M., with Harold P. Green. *The Oppenheimer Case: Security on Trial.* New York: Harper & Row, 1968.

Straffin, Philip D., Jr. "The Prisoner's Dilemma." In *Undergraduate Mathematics and its Applications Project Journal* 1 (March 1980): 102–3.

Strauss, Lewis. *Men and Decisions.* Garden City, N.Y.: Doubleday, 1962.

Teger, Allan I. *Too Much Invested to Quit.* New York: Pergamon Press, 1980.

Teller, Edward, with Allen Brown. *The Legacy of Hiroshima.* Garden City, N.Y.: Doubleday, 1962.

Truman, Harry S. *Years of Trial and Hope.* Garden City, N.Y.: Doubleday, 1956.

Ulam, Stanislaw. *Adventures of a Mathematician.* New York: Scribner's, 1976.

———. "John von Neumann 1903–1957." In *Bulletin of the American Mathematical Society* (May 1958): 1–49.

Vonneuman, Nicholas A. *John Von Neumann as Seen by His Brother.* Meadowbrook, Pa. (P.O. Box 3097, Meadowbrook, PA 19406), 1987.

Williams, J. D. *The Compleat Strategyst.* New York: McGraw-Hill, 1954.

INDEX

About the Author

William Poundstone studied physics at the Massachusetts Institute of Technology. He is the author of *The Recursive Universe,* about information theory and physics, and *Labyrinths of Reason,* an exploration of paradox in science. He is also known as the author of such popular books as *Big Secrets* and *The Ultimate* and of articles for *Esquire, Harper's, SPY,* and other periodicals. He lives in Los Angeles.